a
quiet
tide

a
quiet
tide

Marianne
Lee

NEW ISLAND

A QUIET TIDE
First published in 2020 by
New Island Books
16 Priory Hall Office Park
Stillorgan
County Dublin
Republic of Ireland
www.newisland.ie

Print ISBN: 978-1-84840-754-1
eBook ISBN: 978-1-84840-755-8

British Library Cataloguing in Publication Data. A CIP catalogue record for this
book is available from the British Library.

Typeset by JVR Creative India
Cover images © Zikatuha/Shutterstock.com (front) and
Sneguole/Shutterstock.com (back)
Cover design by Karen Vaughan, www.kvaughan.com
Printed by Scandbook, Sweden

New Island received financial assistance from The Arts Council (An Comhairle
Ealaíon), Dublin, Ireland.

Novel Fair: an Irish Writers Centre initiative

New Island Books is a member of Publishing Ireland.

For J & J

Author's Note

In Ireland in the early 1800s (as in the rest of the United Kingdom) the recipient, not the sender, paid the price of postage. Officers of the state, and others, were permitted to frank and receive an amount of postage duty free.

The binomial system of naming plants that was developed by Carl Linnaeus – the Linnaean system – was introduced in the late 1700s and continues to be used today. However, though the binomial system remains unchanged, the classification of plants has been updated, resulting in name changes for species.

In South-West Ireland in the early 1800s, much of the local population spoke the Irish language, particularly in more remote, rural areas. Many would have had little or no English.

Prologue

Bantry, Co. Cork
February 1807

The fisherman sidled up to her on the quay. 'You're Miss Hutchins, the plant lady from Ballylickey.'

'I am.'

'They said you might be interested in a curious tree I've seen growing.' His English was fluent, if careful. He had steady, intelligent eyes, sand-coloured hair.

'Yes?'

'I know the trees here around, yet I've not seen this type before or since. A rare sight, brave and green, in that barren place.'

'Where?'

He waved a hand out to sea. 'An island, not much more than a big rock, set in rough waters off Bere Island. You'll not get near it except in good conditions, no wind or swell …'

'Could you take me?'

'Too dangerous for a lady.'

'I'm used to boats and climbing to out-of-the-way places.'

'I'd not do it, Miss.' He shook his head, his mouth a hard line. 'But if I get near it again in the summer, I'll dig the roots out for you.'

'Don't put yourself at risk, not for the sake of a plant.'

'I know those waters since I was a boy, Miss, could sail them blind.'

'Well ... I've never encountered anything like you describe. If you get the roots, send them to Ballylickey House.'

She often thought of him, on calm, clear days; she looked for him amongst fishermen unloading mackerel and herring, mending nets, boiling cans of water over small fires. Eventually she approached a group of men congregated on the quays in Bantry. 'Have you seen James O'Sullivan of Berehaven lately?'

They spoke amongst themselves in Irish, gesturing, shaking their heads. Their wind-burnt faces shone like mahogany.

'What is it? What happened?'

One man stepped forward. 'He drowned, Miss. A month back.'

'My God. How terrible. Where?'

They shifted their feet. A low murmur came from within them, like a rushing wave.

'Somewhere off Crow Head. They were only an hour out. A storm came up.' His eyes met hers, a flicker; she saw there all the horror of those few words.

'Had he a family?'

'A wife. Two boys and a girl. He was a good man. They got the body, buried him decent.'

'I'm sorry.'

The men looked at the ground, shrinking within themselves, casting her out. She backed away, shaken, ashamed. Why? She was not responsible. The breeze stung her face, tart with the briny whiff of fish.

James O'Sullivan lay now in Glebe graveyard, under the shifting sky and the whisper of elder saplings.

She searched for the tree, had the crew position the boat as close as possible to every lonely outcrop they came across. She clambered over rock, tearing her skirts. Once she slipped and almost toppled backwards into the sea; she saved herself by grabbing onto some scrubby vegetation. Her hands bore long red scratches, her toe a deep bruise from where she'd pawed for a foothold. A full year for the blackened nail to grow out, sheared away by a gelding knife.

You're Miss Hutchins, the plant lady from Ballylickey.
I am.

DUBLIN, IRELAND
MAY 1803

ONE

Though four storeys of solid brick, Dr Stokes's townhouse gave the impression of lightness.

Wide steps led to the front door. Pillars either side of the door supported a soaring glass arch above. Tall windows reflected scudding cloud. The window rails were of wrought iron, scalloped to look like lace. Standing on the street and tilting back her head, Ellen's eyes flowed upwards into the sky.

Inside the rooms were ordered, the furniture polished and shining. Each room had been decorated with a predominant colour – the hallway red, the drawing room light blue, the dining room green. The large windows and many mirrors threw dazzling light around the upper floors. The ground level was darker and damp but she would have no reason to go there. Unlike at her boarding school – where the few servants scurried non-stop from basement to first floor to second floor, and where the pupils did much of the household chores, dusting, sweeping, making beds – in Dr Stokes's house a different servant appeared for every required task, moving as if oiled, their eyes lowered. A stout manservant opened the front door; a young footman shouldered away her trunk; a thin, pock-faced girl took her bonnet and travelling coat.

Mrs Stokes – Louise – showed her to the room where she was to sleep. The single window faced west and would let in the evening sun. Leaf-green striped wallpaper, a small bed of dark wood, white counterpane and pillows. A rug beside the bed in dark blue and gold; Louise tapped on it with her foot, showing her leather slipper. 'Persian,' she said.

'Very pretty, ma'am.' Ellen said, blinking at its dazzling pattern.

'Come have tea.'

In the drawing room, Dr Stokes stood by the mantelpiece. The manservant was setting out a silver teapot and china cups.

'Alas, I cannot stay, Miss Hutchins. I must leave you in my wife's hands.' His voice held a practised kindness, gentle authority. Despite his long, rather morose countenance, he was fresh-faced, his eyes bright behind heavy lids like a woman's. His hair receded over a high forehead. He wore a coat of black-brown, sober cloth, with plain, silver buttons.

When as a little girl she'd met him, years before, she'd marvelled at his boots, how long and shiny they'd been. And his hand: sinewy and smooth, depositing in her own little paw a thumb-sized piece of sugar. She saved the nugget for days, licking it furtively until it disappeared, to her confusion and sorrow. She'd thought him old. In fact, he must have been in his prime, for he now looked to be in the forties.

He glanced at his wife, as though asking her permission to go.

'We will be quite all right, husband,' Louise said. She sat down on a pale yellow sofa. 'Miss Hutchins and I will become acquainted.'

He closed the door behind him as though afraid of breaking something.

Louise handed a teacup across. 'Do you hear from your brother Emanuel regularly?' She had a serene expression, as though she eked out the movements of her face. Younger than Dr Stokes; red hair under her white cap, blue-green eyes like chinks of glass. So far she had been attentive, gracious, rigidly polite, if unsmiling. She was dressed in dark blue silk; a muslin shawl covered her bosom.

Ellen perched on the edge of a chair, her back straight. 'He's busy in London, practising law.'

'When did you see him last?'

She feared her clumsy fingers might snap off the teacup's pretty, fine handle. 'Some years ago,' she said. He'd visited her in school. She'd been completely overawed by his gruff manner and hadn't managed more than a few stuttered words. He patted her head, called her by her childhood pet name, Ellie.

'How many siblings do you have? My husband couldn't recall ...'

'Four brothers: Emanuel, Arthur, Samuel, Jack.'

'Which one lives with your mother in Cork?'

'Jack,' she said. 'He's ... incapacitated. Due to an accident.'

'You have no sister, then?'

'Katherine died when I was four.'

'Oh. She was what age?'

'Twenty-four.'

'In her prime. A shame.' Louise busied herself with the teapot. She held the spout over Ellen's cup, though Ellen had hardly yet taken a sip.

'You don't like tea?'

It would be rude to say she found it bitter. 'I'm not used to it. We didn't get tea in school. Only milk, or whey. Chocolate, at Christmas.'

'Social life in Dublin, as elsewhere, revolves around tea. It's an essential luxury.' Duly rebuked, Ellen obediently drank. Louise held out a plate. 'Have a sweet biscuit, Miss Hutchins. You're too thin.' Ellen blushed, startled. Louise watched until she'd taken one and nibbled at its edge. 'What was the food like in your school?'

She swallowed. 'Good, ma'am.'

Louise's eyes narrowed to a glint. 'I doubt it. You look half-starved.'

'I've been ill ...'

A colourless woman in a grey dress and long apron entered the room, shepherding three children. Louise gestured: 'Harriet, Minnie, Charles – come closer and meet our guest.' Charles, the baby, fair and red-cheeked, tottered towards his mother. The nursemaid seized Harriet and Minnie's shoulders and steered them forward.

Harriet had Dr Stokes's round chin and protruding lower lip. The hair peeping from under her cap was the same red as her mother's. Ellen guessed her around five years old. Minnie, a year or two younger, stood behind her sister.

'Harriet,' Louise said. 'Take your thumb out of your mouth, you're too old for such behaviour. We will dip your thumb in vinegar, would you like that?'

'No, Mama.' A shining thread of drool had settled on the child's chin.

'Dunne, see to it she stops.' The nursemaid jerked Harriet's arm down. 'Well, then. Miss Hutchins might teach you French if you are good.'

'Certainly I will,' Ellen said.

'Soon a tutor must be arranged. But not yet.' Moments passed. Somewhere in the large room, a clock ticked against the silence. The little girls stared at her with bright, round eyes. Ellen realised she was expected to say something.

'Do you like to draw?' she asked. 'Have you paper, pencils?'

'Of course,' Harriet said, in a clear voice. 'Our papa gives us whatever we ask.'

'Well, then. We can draw together.'

'You are kind, Miss Hutchins,' Louise said. 'But remember, you are here to recover.' The nurse led the children away. Harriet looked back over her shoulder, her fingers already creeping to her mouth. Ellen determined to make friends with all three. Surviving a large household depended on finding allies, from the lowliest members to the highest.

Left alone to rest, she explored the room of striped wallpaper. She couldn't think of it as hers. Not yet.

Pegs on the wall, where her dresses had already been hung, a washstand and basin of white china painted with yellow roses. A chest of drawers, ample, sturdy, with dark oak knobs, smooth to the touch, and – wonder of wonders – a small looking glass. There had been no looking glass at school. Madame Praval said that contemplating your own image led to vanity, and discontent. With something like shock Ellen now regarded herself. The combination of pale skin and fair hair made her look bleached, bloodless; fortunately her blue-grey eyes – the colour of the sea on a cloudy day – had depth, the possibility of sparkle, as when sun breaks on water. She felt along the sharp bones of her cheeks, lifted her chin to study the line of her jaw, fascinated. Then she started and caught herself, moved guiltily away.

The mirror winked light, unblinking as a silver eye.

Her trunk had been unpacked, her petticoats, shawl; the parcel Madame Praval had given her placed neatly in the drawers. 'These are the things you brought with you as a child,' she had said as Ellen stood waiting for Dr Stokes's carriage. 'Take them now. You've outgrown the clothes, of course. They can be sold, or cut up for some useful purpose.'

Now Ellen untied the string and flattened out the paper, revealing two stained and tattered child-sized dresses and a pair of worn sheets wrapped around solid, awkwardly shaped objects. A pewter fork, knife, drinking mug and a silver spoon, that she'd used at every meal for the last thirteen years. She could hardly bring them to table here, in this elegant house where there was surely an abundance of everything. She lifted the spoon, fondled it in her palm.

It came from Ballylickey. Leonora, her mother, had taken it from the dresser, a monumental block of furniture that had occupied Ellen's imagination like a magic mountain. Leonora held the spoon to the fire, her eyes reflecting its shine. Then she muffled it away in a piece of old fabric. 'Leave your other things packed in the bottom of your trunk,' she said. 'But keep this spoon inside your clothes, even when you sleep, until you arrive at school. It's of the best quality.'

The servant Annie spoke from a dark corner. 'The people you'd meet on the road would take the eye out your head.'

'That's enough,' Leonora said. Her fingers plucked at the neck of her dress. 'Stay close to Annie, Ellen, and don't speak to strangers. I wish I could go with you, but I'm not fit for the journey.' She bowed her head for a moment, then grabbed Ellen by the shoulders. 'You know why we're sending you to school?'

'To learn how to be a lady.'

'We advertised for a governess, asked all your father's old friends ... But it seems we're too wild and lonely here for an educated French or English woman to want to live amongst us.' She passed her hand over Ellen's hair. 'Anyway, you should be with girls of your own class.' She lowered her eyes, light sparkling under her pale lashes. 'And there's too much sadness here, for a five-year old child.'

'I understand, Mother.'

She didn't. Why could she not continue to run wild in the fields, around the yard, in and out of the sheds? Always something to discover – puppies tumbling in a basket, a new horse whinnying in the stable. Chirping sounds in the cool, sour-smelling scullery, calling her to a box of flustery yellow chicks. The sea, its shifting colours, how the mountains floated in Bantry Bay like syllabub in sweet wine. On rainy days she climbed into the cupboard in the back hall and hid amongst the mildew-stained cloaks, shut out the rough voices, the slamming doors and stamping boots, the dead pheasants and ducks strung up in the scullery, glassy-eyed, blood crusted on their beaks. When they were home from school and university, Emanuel and Arthur – close enough in age to fight and conspire like twins – swore, knocked her aside, pinched and bruised her with careless fingers. Leonora didn't stand in their way. She stayed mostly in her room with the shutters closed and the curtains drawn. Birthing twenty-one babies – most of whom didn't survive beyond infancy – and early widowhood, had left her wrung out, hollow as a burst seedpod.

Of the journey from Cork to school in Dublin, Ellen remembered little. Annie, muttering – 'Lord save us, Lord save us' – fingering her funny wooden beads, clutching at Ellen for all two hundred miles. The coach hit a rock in the road and she was thrown to the floor, where she lay on her back amidst the shuffling boots of the other travellers, like a flipped over beetle unable to right itself. During a stop along the way she looked out the window and saw a gentleman relieve himself in the hedge, heard a gush of water and grunts of relief. He turned around, shaking drops of water from a fat, pale worm that protruded from his breeches, and smiled at her.

'Poor creature!' her friend Caroline said when she heard these stories. She had never been farther than Clontarf on the coast, and then in her father's jaunting car on a summer's day.

Ellen gathered the old clothes, the sheets and towels, into a bundle. When the servant came she would tell her to take them away. She hid the spoon at the back of a drawer, inside a petticoat. Then she pulled off her slippers and lay on the white counterpane. A new ceiling, with its pattern of fine cracks in the whitewash, little rivers carved across a barren landscape.

It seemed impossible that it had only been the night before that Caroline had entwined her fingers in her own, as they sat on the edge of Ellen's bed in the school dormitory. 'Your last night,' she said. Her voice was bleak. 'So? Are you going home to Cork?'

'No,' Ellen said. 'I'm to live with Dr Whitley Stokes, a friend of my brother Emanuel.'

'Why?'

'My brother cannot come for me.'

Caroline frowned. 'Has Dr Stokes a house in town?'

'Number 16, Harcourt Street.'

Her eyes lit up. 'Close to St. Stephen's Green … When I get home to Henrietta Street, I can visit you. Or you can come to me, if I'm not too unfashionable now you're to live at such an elegant address.'

'I don't care about elegance.'

'Of course, my father might drag me back to his estate in Kildare.' Caroline chewed her lip, thinking. 'Your brother must hope you'll meet a husband in Dr Stokes's house.'

'Who knows? I can't think beyond tomorrow.'

Caroline leaned her temple against Ellen's. 'Will you tell me what happened?'

'When?'

'This morning.'

'Dr Stokes came to visit me. You know I've been unwell.'

'Yes, but why had you to lie down afterwards? Madame warned us not to disturb you, to stay out of the dormitory.' She leaned closer. 'Is it true he went upstairs with you?'

'Madame was there also,' Ellen said.

'Yes, but … did he examine you?'

'I can't remember.' She shook her head. 'I fainted, they said.'

'You fainted.' Caroline sounded breathless.

'Yes, in Madame's study. When I woke I had been carried upstairs.'

'By whom?'

'I don't know. Does it matter?'

'It was Dr Stokes, surely. Madame wouldn't let any of the male servants do it. What happened, after you woke up?'

Madame Praval had stood by the bedside, holding a basin. Ellen stared at the ceiling. A sharp sting, as Dr Stokes drew the blade across her arm. The harsh clatter of drops hitting tin. She heard Madame breathe, '*Mon Dieu*, Doctor.'

'Beneficial, I assure you, Madame.' Ellen focused her gaze on the creases around his eyes, the whorls of hair in his ears. Afterwards he bound the wound with boiled rags. It throbbed still, under her sleeve.

'Nothing happened,' she said. 'He asked me some questions, laid a cold cloth on my forehead.'

'Oh.' The light leaked out of Caroline's eyes. As always, she burned for the out-of-ordinary, the potential for drama, however slight, her defiance against the mundane. She shrugged, squeezed Ellen's fingers. 'Well. We must make the best of our lot, I suppose, I without a mother and you without a father. At least it's almost over.'

'What is?'

'This waiting, for life to begin.'

In the strange room of striped walls, Ellen rubbed away the beginnings of tears. Weak. Ungrateful. Imagine Caroline, instead: here, with her as she had almost always been. Describe for her Louise Stokes, her blue dress, red hair and precise, cool speech. The perfect, pale yellow of the sofa, like sun on a frosty morning.

∽

'Miss Hutchins.' Dr Stokes's voice. 'Are you awake?' She crossed to the door and opened it. He stepped back, cleared his throat. 'Ah, you're up.'

'I was resting, sir.'

His eyes flicked over her shoulder to the rumpled bedclothes. 'Very good. You need rest, and quiet, until you feel stronger.' He peered at her. 'It also occurred to me that you might need amusement.'

'I'm content, sir, thank you.'

'Nevertheless, distraction is good. Do you like to read?'

'Yes, sir.'

'Come with me.' She followed him downstairs, down the corridor, past the drawing room. He stopped at a closed door and produced a key from his pocket, which he wagged at her. 'To keep the children out. You may ask for it whenever you please.'

Behind the door, a large, dim room. A desk covered in papers, two easy chairs. She squinted upwards. The shelves were packed with books, the titles blurring as they rose higher.

'Had you access to books in school?' Dr Stokes asked.

'Much of the library was falling to dust, eaten by moths and silverfish. Old French literature and poetry, mostly.'

He had taken a book down from the shelf and caressed the binding. 'I take precautions against such damage,' he said. 'It's essential to have the room aired and dusted, lay arsenic if necessary.'

'Yes, sir.'

'Now, literature, this side – poetry farther down, by author, of course – science over here – consult with me first, that subject is so broad … have you any interest in science?'

'I cannot say.'

His face creased, surprised and amused. Had she been impertinent? 'What did you study, all those years at school?'

'French, of course. Drawing, sewing, dancing, literature. Some geography …'

'Latin? Mathematics?'

'No, sir.'

'None of the natural sciences? Biology? Botany?'

'No, sir.' She looked at the floor. 'Madame Praval said young ladies should to be taught to be useful.'

He grunted. 'Whatever that means. I suppose half an education is better than none.'

'I like to draw flowers,' she said. How limp, how girlish, that sounded.

'Well, that's a start. Drawing is a most suitable pursuit for a lady. Are you proficient?'

'Tolerably, sir.' Bolder, 'It was my best subject.'

'You'll find illustrations here worthy of study.' He ran his finger along the row. 'And I have some fine volumes on antiquities. Ah!' He found what he was looking for, plucked it from the shelf, tapped the cover. 'Here it is. Rousseau's *Letters on the Elements of Botany, Addressed to a Lady*. I purchased this translation for Harriet to read later on. But why preserve it until then? You are his precise intended audience. Have you read him before?'

'Yes, sir.'

'Of course, you'll have read all the great writers in French … but not Rousseau's treatise on botany? Well, let us speak when you've read it.' He paused. 'You've no objection to my recommending books to you?'

She blushed. 'I'm obliged to you, sir.'

He waved this away. 'If you're interested, there's much here for you to discover. Unless you're content to sit and stitch all day like my wife …' He stopped, frowned. 'Like other ladies.'

'My needlework is so poor, sir, I fear there would be no benefit to anyone if I did. I never acquired the skill, no matter how many hours I cramped my neck over it.'

'No one will force you to sew anything here.' He pulled out another volume. 'Mrs Wakefield's *An Introduction to Botany*. Interesting to read a female perspective. But start with Rousseau, I say. If nothing else, his expression is elegant, his theme diverting and eminently suitable.'

That word again. Always a judgement. Or its opposite, *unsuitable*, a condemnation. *Look away, Miss. Danger here*. 'It's a wonderful library, sir.'

He shrugged. 'Of course, it's not always possible to get the latest publications in Dublin, but I have a seller in London who can source particular volumes if needs be.' He placed the key on the desk. 'I'll leave you to make your own discoveries. There are subjects other than botany, after all. My own particular passion: I'm inclined to proselytise. Forgive me.' He peered at her. 'Your colour is suddenly high, Miss Hutchins. Are you tired? Overly excited?'

'No, sir.' Not tired. A strange, not unpleasant feeling of blood bubbling through her veins and rushing in her ears.

'I want to take your pulse, Miss Hutchins,' he said. 'Do you remember when I did so, yesterday?'

'No, sir.'

'Give me your arm.' He pulled a watch from the fob in his breeches and clasped her wrist between his thumb and third finger. His head bent towards hers. Like a sudden ringing noise, she became aware of his smell, a peppery tang of soap and wool. Warmth and a numbing tingle spread along her arm. He stared at the watch, increased the pressure of his fingers.

'What is it, sir?'

He let go of her wrist. 'Sit, my dear. I'll ring for water. You're still weak.'

Later, helped back to her room and once again lying down, she opened the book he'd given her: *The principal misfortune of Botany is, that from its very birth it has been looked upon merely as a part of medicine. This was the reason why every body was employed in finding or supposing virtues in plants, whilst the knowledge of plants themselves was totally neglected: for how could the same man make such long and repeated excursions as so extensive a study demands; and at the same time apply himself to the sedentary labours both of the laboratory, and attendance upon the sick ...*

She tossed all night on the cushioned mattress, under cotton sheets; she had become used to thin bedding, rough blankets. Her mind played over the last moments of her schooldays.

They had filled the front hall, a huddle of pale-faced girls: tall, thin, small, dimpled, standing straight as rush-lights, hands neatly clasped. When they heard the sound of wheels crunching on gravel, the younger girls began to sniffle. Caroline made no effort to wipe her eyes, catching the tears on her tongue as they dribbled past her mouth. The tightness in Ellen's throat made it difficult to swallow. She blinked furiously, looking upwards.

'Stand up straight, Mademoiselle Hutchins,' Madame Praval said. 'You will not make your height any less, only appear as if you are hunchbacked.' With her sharp nose, tightly wound hair and dark dress she resembled a watchful crow. That morning she'd called Ellen to her study. 'I never asked you to leave.'

'No, Madame. But my fees have not been paid this quarter. Twenty pounds, you said.'

'Dr Stokes has promised to settle them.'

'I've lived on your goodwill long enough. It seems I must now live on someone else's.'

'You're content with this arrangement?'

'Content to do as I am bid, Madame,' she said. 'I will try and make myself agreeable.'

'You have always been that. Agreeable, if unremarkable, without a defined personality. I say this to your credit.' Madame's eyes, dry, black as coal, fixed her like an insect under a glass. 'You know, I have tended you as if you were my own child. No mother could have done more.'

'I know, Madame.'

'I hope you will think on that, and speak well of us when you've gone.'

'Yes, Madame.'

The lines at the corner of her mouth flexed and softened. '*Bien.*' She stood from her desk, came close to Ellen and stretched to rap softly against her forehead with her knuckles.

'Somewhere in here, there is confusion. I see it.' Ellen said
nothing. Madame nodded. 'You know what I mean. You must
strive to curb it. Accept what is around you. Do not fight.'
Her breath smelled of coffee and warm decay. 'You will wear
yourself out, you will fade to nothing, Mademoiselle. Quietly,
politely. And no one will notice.' Ellen's eyes misted; Madame's
face blurred to a pale cloud. Outside in the garden, a blackbird
sang. 'So, you won't speak. But I know you understand me.
With those searching eyes, you see all.'

In 16 Harcourt Street, Ellen turned over once more on the fat
mattress, listening in vain for the sound of familiar breathing
in the dark. With no one to witness, she put her thumb in her
mouth and held it there, wondering if it might bring comfort.

TWO

'She barely eats,' Louise said, frowning as she tugged a thread through the silk.

'Who?' Dr Stokes held a book in one hand and absent-mindedly toyed with the paper knife with the other.

'Miss Hutchins, of course.' Louise used her voice as a musician might an instrument, accelerating, pausing, cutting from legato to staccato; she played the pianoforte in the evenings – hymns, sentimental ballads remembered from her girlhood. 'She crumples a bit of bread, sips at a cup of whey. Her hands shake, her teacup rattles in the saucer. Her wrists are mere twigs.' The needle flashed in, out, in, out. 'I know it's in fashion for young ladies to pick at their food ...' Dr Stokes raised his eyebrows. 'It's true, I heard it from Mrs Langley, she despairs of her own daughters. But Miss Hutchins seems so modest. There's nothing in the least fashionable about her.'

'She's not been nourished properly. I suspect an underlying ailment.'

'Something serious?'

'A weakness of the nervous system, maybe. She's highly strung, like many young women. She may grow out of it.' He closed the book. 'I think she should stay with us a while longer.'

Louise's brow puckered. 'Is that necessary? Would she not fare better with her own family? Her mother?'

'I don't think she would survive the journey to Cork.'

She stared at him. 'As ill as that?'

'I fear so. The roads, even now with the fine weather – the pestilent inns, the ill-cooked food ... she isn't strong enough to endure it.'

'This household is full as it is. I don't have time to nurse an invalid.' Her face remained calm, but he recognised the rigid set of her shoulders.

'I couldn't leave her in that place, in that condition,' he said. 'She doesn't require nursing, simply rest and comfort. We can offer her regular family life. She hasn't known much of it. The headmistress at the French school – a strange woman called Praval – says she's quiet and respectful.' Still she frowned. His final effort: 'She's used to children and could help with the little ones. Didn't you say she's offered to draw with the girls?'

She pursed her lips, her mind ticking over. Finally, primly, she said, 'God has directed her, through your intercession, to our care.'

'Her mother will be grateful.'

'And her brother?'

'Of course.' She said nothing, smoothed her hand over the taut fabric. Louise thought ill of Emanuel Hutchins. Dr Stokes had not shown her Hutchins's letter.

... I have not yet decided what my sister Ellen's immediate future is to be. As you know, our father died when she was two years old; thus the responsibility falls to me. Though my mother is desirous of her company, I wonder as to the wisdom of dispatching her to our lands in Cork so soon. She is more likely to form advantageous connections in Dublin. Besides, I am told she is delicate. Could I presume upon both your medical expertise and your judgement by asking you to visit her on my behalf? I can't leave London at present and my mother could not possibly make so long and arduous a journey. I will take your advice on the matter. She could, perhaps, make a useful companion to some lady of your acquaintance ... As for me, I go along here tolerably well. As disenchanted as I was with Ireland

at the end, now that I am away I am torn between despair and affection for the place. I am destined to never feel completely at ease wherever I go. When I was in Dublin, I longed for Cork and our estates. Do you ever see any of our old friends – those that survived the madness? Your obliging friend, Emanuel Hutchins.

Acknowledging his responsibilities as the head of the family, at least. Yet also a touch of the old brooding introspection. Five years before, Dr Stokes had feared for Emanuel Hutchins's sanity. When last they'd met he'd been seeking comfort at the bottom of a whiskey bottle. Now he appeared to have settled: a man of the law, of all things. How long would it last?

He sighed. 'I thought the girl might have recovered some of her appetite by now. The food here is surely preferable to what she got before.'

Louise clicked her tongue. 'Molloy is offended, she's taken great pains. She ordered chicken particularly. You must tell the girl to eat, sir.'

He leaned his head back. The curtains hadn't yet been shut; he could see into part of the drawing room opposite. An elegantly dressed woman stood near the window. Her lips moved, she laughed and shook her head at someone out of view. A play performance, without words.

'If one is queasy, or unwell, an abundance of food is overwhelming,' he said. 'She's not used to jellied pigeon and sweetbreads. Have some morsels placed in her room. Not a tray, but a choice dish of something light, appetising. Tell the servant not to speak of it, but to take it away if uneaten. And tell her she may breakfast and dine in her own room, if she prefers. Let her choose to come to table.'

'You think that will make her eat? I won't have it said that a young woman starves in my house.' Her voice had a hard, silvery edge. 'We may have to send her back to the Frenchwoman.'

'No,' he said. 'She just needs time to adjust to her new surroundings. We must lead her, as a bird is led by a trail of crumbs into a garden.'

Louise's lips twitched. 'If you can get the bird to eat the crumbs, that is.'

The rumple of curtains being drawn, a soft voice. 'Will I open the shutters too, Miss?'

Ellen murmured 'yes'. Clanging wood and metal, a flood of harsh light. She put her arm across her eyes.

'There's fresh water in the bowl,' the servant said. 'I'll come back in a minute.' The door closed.

Ellen got up and pulled the chamber pot from under the bed, squatted over it to pass water. Afterwards she washed her hands. The water was warm and the soap had a faint floral smell – not roses, less sweet, woodier. She unbuttoned her nightdress, pulled on her drawers, fastened her stays under her breastbone.

There was a small bowl on the table that had not been there before.

The servant must have brought it. She padded across the room, the floorboards cool and pleasantly rough against the soles of her feet. She lifted the bowl to her face: cream, and a handful of red berries. She looked at the door. Her stomach grumbled. She dipped her finger in the cream, sucked, tasted sugar. Her tongue curled and saliva flooded her mouth. She used the berries to scoop up the cream, the pips cracking against her molars. When she'd finished, she wiped her mouth and hands. Then she took down her Sunday dress, her least shabby. She'd half pulled it on, folds of material pooling around her legs, when the servant knocked again.

It was the young, pock-faced girl, drab and tidy in a brown dress and cap. 'Let me, Miss.' At school they had helped each other dress, turning to offer their backs, raising their arms as if in a well-rehearsed dance.

The servant's fingers tugged at the buttons: a pleasing sensation. 'The tall trees I see from the drawing room window,' Ellen said. 'Are they part of a park?'

'That's Clonmell Lawns, Miss, that belong to the Earl.'

'Can I walk there?'

'Well, no, Miss. The gardens are for the Earl's private use. There's an underground passage from his house.'

'I see.'

'There's a garden at the back here, Miss.' She'd seen it from her window: cramped, covered in shadow most of the day. 'St. Stephen's Green down the road,' the girl went on, her breath on Ellen's neck. 'And the master often goes on trips to the country, from dawn 'til last light. Sometimes he brings Mrs Stokes and the children with him for an afternoon. Collecting plants, he is.'

'What for?'

The fingers paused. 'I don't know exactly, Miss. He keeps them, I think, in a great book. Very interested in plants, is the master.' Now the fingers continued, down to the base of Ellen's spine. 'There, that's done. Will there be anything else, Miss?'

'No, thank you. Janey, isn't it?'

'Yes, Miss.' She showed a glimpse of jagged teeth and lifted the bowl from the table with a questioning look.

'I've finished,' Ellen said. 'Wherever did they get those berries?'

'Raspberries, they're called, Miss. Fresh from Lord Clonmell's garden, especially for you.' She reddened, bobbed and fled.

The next morning she left a dish of custard, dusted with a heady, rust-coloured powder she said was called cinnamon. Later in the day, coming back to the room to rest, Ellen found two potato cakes on the table – still warm, wrapped in a napkin – and a thimble-sized glass of sweet wine. She ate the cakes with her fingers, licking the warm fat – butter? – that trickled down her wrist. They had been given lard in school. Somewhere within her the taste of the butter resonated, like the ringing of a distant bell.

She ventured later into the drawing room, sat by the window watching the passersby on the pavement opposite. Louise came in.

'Ah! You've joined us.' She seemed surprised, but pleased.

Ellen stood up. 'I wanted to thank you, ma'am. For the delicacies you had brought to my room.'

'Dr Stokes thought it best. I must say, we racked our brains to think of something suitable.'

'They were delicious. You're very kind.' She lowered her eyes, away from Louise's razor-sharp glance, endeavoured to keep her voice humble and grateful.

'Well. No doubt you're an obedient young woman. Do as the doctor tells you and you'll soon have roses instead of ashes in your cheeks.'

The children moved about the upstairs rooms like ghosts, tiptoed on the stairs, suppressed their shrieking laughter. Whispered outside her door, daring each other to call her name – 'Miss Hutchins? Are you there, Miss Hutchins?' Then soft giggles, scuffling, the sound of slippers brushing the wooden floor as they fled. She turned her head and dozed, or sat at her window drawing faces from memory, the cloud shapes passing over. She grew used to the idea of ownership, of privacy, the privilege of her own door, a key turned in the lock.

She gained strength. With Louise's agreement she coaxed the children from their sneaking games, gathered them in the nursery for an hour one morning. She placed an apple in the centre of a high table. The nursemaid, Dunne, watched; as soon as Ellen came into the room she'd taken a chair, placed it in the corner and arranged herself on it neatly, hands folded in her lap. She had a pale, pasty face and a sulky mouth; she might have appeared dull-witted if not for her staring, pale-blue eyes. Ellen turned so that she might not see her.

'Observe the apple,' Ellen said. The little girls looked at her blankly. She raised her voice. 'Is it absolutely round? How shall we suggest the curve of its surface? How shall we give it weight? Watch how I do it.' She quickly sketched the fruit while they watched, Harriet attentively, Minnie fidgeting, eyes roving the room for escape.

'It looks real,' Harriet said. In fact, the drawing was clumsily, hastily executed; children were easily impressed.

'And we haven't yet added colour. Now, your turn.' She helped each of them mount paper on a drawing board. 'Begin.' Silence, except for the scratching of their pencils and small sighs from Minnie. 'Come, you may enjoy it,' Ellen said. She leaned over Minnie's small shoulder, catching her child's smell of powdery scalp. 'I would like to tell your Mama that you made an effort.'

'I can't do it.'

'How can you tell, if you give up so easily?' She pressed the pencil into her hand. Minnie pushed out her lower lip, allowed Ellen's hand to lead her initial strokes. 'See? You've made a start.' She walked between them, correcting, encouraging. 'You must follow the line, Harriet.'

'I am.'

'It must be as it is, not what you imagine it should be.'

'I'm trying.' Harriet frowned with concentration, her tongue curling from the corner of her mouth.

After fifteen minutes Ellen said, 'Now, we start again.' They looked at her, open-mouthed, as she unpinned their crude efforts and turned the sheets over. 'Begin.'

Twice more, she made them restart, ignoring their complaints. 'Better,' she said. 'You begin to see.' She paused in front of Minnie's latest attempt. 'You have a natural hand, Minnie. Well done.'

The door opened and Louise came in with a swish of skirts. Harriet sat straighter, set her face to amiable sweetness. Minnie put her pencil down.

'Well?' Louise asked. 'How have they behaved, Miss Hutchins?'

'Admirably.'

'Good. May I see the results?' She walked from one to the other, made approving sounds. 'Minnie, your drawing shows promise.'

'I agree,' Ellen said. 'She has talent, if she perseveres.'

'And Harriet?' The child lowered her eyes.

'Harriet works hard,' Ellen said. 'And will also improve, over time.'

'It seems for once you have done less well than your younger sister,' Louise said. 'It can only encourage you to try harder. Well, that's enough for one day, I think. Time for a walk with Dunne.' Minnie jumped up immediately and ran to the nursemaid, who began levering herself from her chair. 'Come, Harriet,' Louise said, 'Put your drawing things away.'

She reluctantly laid down the pencil, pushed back her chair. Ellen caught her eye, smiled. The little girl's returning stare was shockingly adult in its cold dislike.

⁓

September: a last blast of warm weather. A sand-like dust from the street infiltrated the front rooms, even with the windows closed, and was carried on slipper soles and hands to the rest of the house. Louise scolded the servants, despairing over her wallpaper and upholsteries. The children complained of ticklish throats, stinging eyes.

For respite, Dr Stokes promised a Sunday outing.

After early service in St. Andrew's, they prepared to head away. First a fuss over who would sit where – who would take the baby, the picnic; at last Dr Stokes, Minnie and Harriet climbed into the first carriage, Louise, Ellen, Charles, Dunne and a young manservant, Peter, into the second. From the church, they turned for College Green. Ellen had a new bonnet of straw that blocked her view to the left and right like the blinkers of a horse. She strained her neck as far as she could. Fine buildings lined the wide streets. Well-dressed gentlemen and ladies, some garish in bright colours, strolled the paths. As soon as the carriages slowed, ragged men, women and children materialised from the alleys, holding out dirt-caked, pleading hands. 'Please, young Miss, we're starving …'

'Get away!' Peter raised a stick. The driver urged the horse on. The last thing Ellen saw as they lurched forward was a tiny infant, no more than a scrap, clinging to its mother's ragged bosom. She shrank back, glad for once of the curved wings that shielded her face.

From the centre to the outskirts. The fine buildings thinned out, replaced by desolate spaces, cobbled or covered with patchy grass. Dogs and pigs roamed, urinating and defecating freely. Dunghills steamed in the sun. Louise held the end of her shawl over her face, and Ellen did the same, the stench overwhelming. Now they passed a row of cottages with well-tended front gardens and lines of washing flapping in the breeze. Barefoot children scampered about. Soon the straggling outskirts merged into countryside and thick hedgerows lined the road, high with grasses and poppies, Queen Anne's lace, oxeye daisies. The horses clipped on at a steady pace through clouds of midges and horseflies.

'They're drawn by the horses' sweat, ma'am,' Peter said. 'Sweet as sugar it is to them.' His voice was respectful, his quick glance at Louise furtive and sly. She seemed not to hear him, waving her hand in front of her face. Moisture coated her upper lip. Damp patches had spread under Ellen's armpits; her thighs itched against the leather seat.

'The sea!' Harriet stood in the leading carriage, gawking for a better view and holding on to her straw hat, almost falling out as the carriage rocked over a rut in the road. Dr Stokes pulled her back down. 'The sea, the sea!' Minnie echoed. The road now clung to the coast and the great curve of Dublin Bay came into view. In the distance, low hazy hills covered in yellow and purple vegetation fell towards the sea. Across the bay, tall ships streamed into Dublin harbour, white sails stark against the blue, while fishing boats bobbed about closer to shore. The sea churned with breakers in the stiff breeze. At the village of Clontarf, no more than a shambling cluster of cottages, bathing machines had been pushed into the tide, though they didn't appear to be in use. An old man sucking on a pipe waved from his seat on the sea wall.

'Lovely, is it not?' Louise sounded half asleep.

'I was born beside the sea,' Ellen said.

'Yes. But you're more used to city ways now.'

'I know nothing of "city ways".'

'We must take you out more. You're young, and should meet other young people.'

Ellen had a sudden vision of being delivered from salon to salon to drink tea and gossip. She would stutter hopelessly; she had none of Caroline's talent for chatter. And her clothes: plain, without lace, ribbons, beading, or embroidery … 'I prefer quiet occupations,' she said.

'Too quiet,' Louise said. 'Dr Stokes is impressed with how you're working your way through his library. But think, when I was your age, I enjoyed going out in society. Where else would I have met Dr Stokes?' She fanned herself with the edge of her shawl. 'Though I little thought then that I'd have to compete with plants for a husband's attention.'

'How long has Dr Stokes been interested in botany?'

'Since he was a boy. But I would not refer to it as an interest, rather an obsession, such as men frequently have. Horses, music, Italian poetry, naval battles ... They apply their fervent minds to something other than the everyday business of living and dying, which is all women have to occupy them.'

'Perhaps because of his profession Dr Stokes takes comfort in all that's green and alive,' Ellen said.

Louise gave her one of her looks, not quite affronted, alert to the possibility of affront. 'He plucks plants from the fields and forests where God placed them, crushes them into dry scraps. Stares at them for hours under his microscope as if they held the meaning of life itself.'

'Rousseau says that nature ...' Louise's nostrils flared, a warning. Too late to stop: 'He says that nature *abates the taste for frivolous amusements, prevents the tumult of the passions, and provides the mind with a nourishment which is salutary, by filling it with an object most worthy of its contemplations.*' She blushed. 'Or words to that effect.'

'My mind is already full of objects "worthy of my contemplations",' Louise said. 'The dinner menu. Charles's rash. The need for a new butcher.'

'Of course, ma'am. It must be difficult.'

'I suppose you can't be blamed for being young and naïve. You'll soon learn that books will not help you escape life. Mine has been hard but I don't complain. What's the use? What cannot be changed must be borne, with grace.'

Ellen thought: life, with nothing beyond domestic cares, would soon prove intolerably dull, for man or woman. Louise at least had her cosseted children. She said none of this, however, only nodded in sympathy.

Outside the fishing village of Howth, just as the incline became too steep for the horses, the carriages stopped. Peter jumped down and gave his hand to Louise, then Ellen. They walked, all of them, for a time, the servants trailing after with the baskets. Eventually Dr Stokes called halt. A magnificent sweep of blue, the green-gold coast. Gorse, alive with the thrum of bees, covered the hill in a fug of sweetness.

'Will you play cards with me later, Miss Hutchins?' Harriet stood at her elbow, squinting upwards.

'Will you not play with your sister?' Ellen said.

'She has no patience and will not learn.'

'Then I will, with pleasure.'

Harriet smiled, satisfied. Ellen said, 'Is the view not very fine? Though no doubt you've seen it before.'

The little girl shrugged: perhaps. Then slowly, lightly, as though testing the air, she said, 'I heard Dunne say the English almost hanged your brother.'

Ellen blinked back shock. 'To whom did she say this?'

Harriet fluttered her lashes innocently. 'I can't remember.'

'Dunne has been listening to gossip and lies. She should know better.'

'Oh.' Harriet looked nonplussed. 'She also said you're dying. That you'd not live past Christmas.'

The line of blue horizon blurred and tilted. She made her voice firm. 'I was unwell, but have recovered.' She grabbed the child's arm, squeezed it. 'Is my grip not steady?' Confusion came into Harriet's eyes, and the beginnings of fear. 'Well?' Ellen squeezed tighter.

'That hurts, Miss Hutchins.'

She held on for another few seconds, before letting go. 'You see I am quite strong.' Harriet rubbed at her arm, where the imprint of Ellen's fingers showed pink. 'You can tell Dunne.'

'Come and have some refreshment, Miss Hutchins!' Louise gestured to the spread out blankets on the grass. Peter was laying out the family's picnic: a pie of cold tongue, fruit and sweetmeats. Harriet gave Ellen a reproachful look

– though was there also a tinge of guilt? – before running to throw herself down close to her mother. Dunne trailed Charles, who had made a dash for freedom, flapping in his white baby skirts. Dr Stokes held a glass of wine to the sunlight. It glowed crimson against his hand.

❧

'I'm going to walk across the headland,' Dr Stokes said. He hung a flattened cylindrical case across his shoulder, suspended from a strap. 'My vasculum,' he told Ellen. He squatted down beside her and unhinged a lid in the centre of the case. 'Where plant samples are stored.' He allowed her open and close the lid.

'Ingenious.'

He stood and pulled his hat on against the breeze. 'Later I will show you how a microscope works,' he said, 'The window to another world.' He strode off. Minnie and Harriet followed him for a distance; he turned and directed them back to where Louise sat on the grass.

'Could they not accompany him?' Ellen asked.

'In that case we should all have to go,' said Louise. 'He's distracted when collecting plants. The children would end up tumbling off a cliff. The way he scrambles down to the shingle, over rocks, through the gorse – we could not manage in our skirts without getting caught on thorns or tripping over our petticoats. We'll take a walk that way' – she indicated a flatter part of the headland – 'and after that Harriet wants to play cards.'

As they walked, Ellen taught the children a nursery rhyme.

À Rouen, à Rouen, sur un petit cheval blanc
À Verdun, à Verdun, sur un petit cheval brun
À Cambrai, à Cambrai, sur un petit cheval bai

Revenons au manoir sur un petit cheval noir
Au pas! Au pas!
Au trot! Au trot!
Au galop! Au galop!

Minnie would only make foolish noises and run ahead when reprimanded. Harriet quickly memorised the rhyme and repeated it back perfectly in a lisping singsong. Louise clapped, and nodded at Ellen approvingly. Harriet was clever, Ellen thought, like her father. Though with a different nature. She checked herself. Be fair, she thought. She's only a child.

Tittle-tattle. Pointless, vicious. Caroline always said that men's indiscretions ended with duelling pistols, while women cut at each other with their tongues, over tea and sweetcake.

By a rocky outcrop resembling an old man's profile, she stooped to pick a few of the last wild flowers that clung on in the unseasonal warmth. Pressed between the pages of a heavy book, they would retain a faded colour, memories of themselves. She walked on, eyes skipping past daisies, clover, herb Robert … But what were these? – small, mauve-lavender flowers, scattered like stars amongst a patch of low-growing leaves, not like anything she recognised. She kneeled and plucked a sample, releasing a sweet, heady scent. The stalks left a sticky residue on her gloves. She folded it in her hand-kerchief with the rest.

Back at the spread-out blankets, she placed the samples carefully between the pages of her book.

'You wanted to play cards, Harriet?'

'Oh yes, please, Miss Hutchins.'

Harriet dealt the cards, singing to herself, snatches of the French rhyme. The pinch had faded from her arm; she seemed to have forgotten, or forgiven, what had passed between them. Fickle, as the white butterflies that flittered over the gorse.

Minnie shouted, 'Father!'

'At last. Peter, pack up, we'll leave now for home.' Louise began wrapping up her needle and cloth. 'Were you successful?' she asked as Dr Stokes walked up, panting.

'Moderately,' he said. 'I found one or two specimens of interest.'

Louise raised her eyebrows, tilted her head. 'It seems you have a rival.'

'Miss Hutchins? Have you been botanising?'

'Nothing so organised, sir. I've collected some flowers for my notebook, that's all.'

'The variety of wildflower for the time of year is remarkable.'

'I found one that seemed unusual,' she said.

'Ah yes?' A twitch of interest. 'Show me.'

'I'm sure it's nothing very particular,' she said, opening her book.

'May I?' He pulled his eyeglasses from his coat pocket. 'Interesting,' he said.

'What is it?'

'*Erodium moschatum*. Musk stork's-bill. Or so I believe, I've never seen it before. Can you remember where you found it? Locate the spot?'

'I think so.'

'Guide me there. This way?' He started in the direction they'd taken their walk, gesturing at her to follow.

'Dr Stokes!' Louise said. 'The children are tired and hungry, as am I.'

'We'll be no more than ten minutes. Then we'll leave promptly, I promise.'

They took off along the path, Ellen retracing her steps, looking about keenly – had she missed it? No – relief! – here it was, behind the curious rock. Dr Stokes murmured to himself as he took a knife from his knapsack. She kneeled to watch him. 'We take the roots, if possible,' he said. 'But there is so little of it here that I'm reluctant to do so.' With a single cut he sliced the stem low, stood and examined the cutting. 'Unusual, indeed. You have a keen eye, Miss Hutchins.'

'I can see objects close to the ground well enough, sir. Beyond that everything begins to blur. I fear I am becoming increasingly short-sighted.'

'Finding plants is not a matter of eyesight,' Dr Stokes said. 'It's an instinct.' He snapped shut the vasculum. 'Come, we'd better get back before Mrs Stokes leaves without us.' Ellen hurried to keep pace with his long stride.

'Here we are, Plunkett, in need of hot water to wash off the dust, and our dinner.'

The servant stood aside as they crowded into the hall, a jumble of children, baskets, shawls and hats. As Ellen passed, he cleared his throat. 'This came earlier, Miss,' he said, handing her a letter.

'Whomever can that be from?' Louise asked. Ellen held the letter behind her back, like a child unwilling to share. As soon as she could, she escaped to her room, shut the door, leaned against it for a moment. Blessed quiet. Then, a low buzzing, from somewhere in the room: she searched, found a bee crawling across the window glass, oblivious to the couple of inches of open sash beneath it. With the edge of the letter, she carefully flipped it outside. It staggered in loops, distrusting its freedom, before spinning away into the air.

Dear Madam,

I am studying at Trinity College where I recently had the honour of meeting Dr Stokes; hearing that I am from the southwest, he asked if I were acquainted with your family. You can imagine my astonishment when he told me that you are currently residing in his house. I believe my maternal

great-grandmother and your great-grandmother were sisters.
I have relatives living at Inchilough, very near Ballylickey. So
you see we are distant cousins as well as having a neighbourly
connection. I hope to call on you at Dr Stokes's house tomorrow
evening, for I'd very much like to make your acquaintance.
Exiled as we both are, we should be friends.

Your obliging servant,
Thomas Taylor

She read the letter aloud at dinner. 'It slipped my mind
entirely,' Dr Stokes said. 'He's an attractive young man of
around twenty, though I can vouch there's no family resem-
blance. He's as dark as you are fair. We should have him to
dinner. Students are always in want of a good meal.'

'I doubt Mr Taylor expects as much as that,' Ellen said.

'On the contrary, I'm sure he's counting on it. We've fed
many of his ilk. They come under the guise of discussing some
point of their education, when what they actually want is a
slice of mutton and a glass of good claret. What do you think,
Mrs Stokes?'

Louise dabbed at her lips. 'I would be delighted to meet
one of Miss Hutchins's relations. However obscure.'

'Hardly obscure,' Dr Stokes said. 'A cousin, he says.'

'Did your mother never mention a relative named Taylor?'

'No, ma'am. Of course, I have not spoken to my mother in
many years, except through letters.' These had come sporadi-
cally; regularly, at first, before dwindling away on both sides.

Dear Mother, my handwriting is improving. I have outgrown my
shoes; please send money for new ones. One of the older girls has been ill
for a month, they have sent for her father.

Dear Daughter, a fox got in the henhouse and killed every one of
the chickens. Your brother Jack manages now with a stick. Wear wool
next to your chest to prevent the ague.

She still received a note and parcel at Christmas: *Thinking of you, dear daughter, at this time of year.*

The thought struck her with a chill: would her mother know to send it to Harcourt Street, presuming, that is, she was still here?

'Miss Hutchins.' Ellen looked up.

'I was saying. By all means, ask your cousin to dinner.'

'Thank you, Mrs Stokes,' Ellen said.

'It's a rare evening we don't have guests, after all.'

'That's settled,' Dr Stokes said. His chair scraped backwards as he stood. 'Come, Miss Hutchins, I promised to show you the microscope.'

In the library, he indicated for her to sit at the desk. 'I prepared the specimen you found earlier,' he said. 'See how it is transformed.' He pulled a black leather cover from an object the size of a large water-jug, revealing a shining metal apparatus. He moved it to the centre of the desk. 'Apply your eye to the lens.' The metal felt cold, clamped to her eye socket. He guided her hand to a brass screw. 'Adjust like so, until the specimen comes into focus.' Something within the apparatus shifted, the blur of colour separated and clarified.

'I see it.'

'Yes?'

She drew back, excited, placed her eye to the lens again. 'Is that the cutting you took?'

'What do you see?'

'The back of the flower ...'

'Calyx. And here ...' An implement appeared under the glass, many times magnified, turning the flower, probing its centre. 'Petals, stamen, carpel, anthers.'

She stood back, looked at him. 'To think this exists, beyond human perception.'

'Wonderful, is it not?' He smiled. 'I told you, another world.'

Later, she went to her room to fetch her book. When she came back into the drawing room, Dr Stokes and Louise were standing close together beside the fireplace. His hands encircled her waist. When he saw Ellen he dropped them and sat down, shaking out his newspaper and raising it in front of his face. Louise fussed with the candle on the mantelpiece. Ellen opened her book, cursing the blush she could not contain no matter how she tried.

◈

If Mr Taylor was awed, he hid it well. He bowed and smiled: a tall young man, compact rather than slim, and, as Dr Stokes had suggested, exceedingly dark for an Irishman. His hair shone like the sleek pelt of some water-creature. Louise and Ellen stood to greet him. Dr Stokes hadn't yet returned from his rooms on French Street, where he saw patients.

'My husband was correct, Mr Taylor,' Louise said. 'You look nothing like your cousin.'

'I take after my mother, an Indian lady of rank,' he said. 'High-born, in that culture.'

'Indian!'

'Yes. I was born in a boat on the Ganges.'

'Remarkable.' Louise seemed entranced. 'And your father?'

'An Irishman, a captain in the artillery. He sent me back to be schooled when I was seven.'

'Alone!'

'Certainly.'

'What an adventure for a boy.'

'When I wasn't numb with terror, I fancied myself in a story-book.'

'An experience like that must have a profound effect on a child.'

'It made me independent from an early age.'

Louise nodded towards Ellen. 'Miss Hutchins,' she said, 'was also sent to school young.'

'Yes,' Ellen said. 'At five.'

Louise sniffed. 'I can't imagine sending Harriet away.'

'My parents set sentiment aside for the sake of my education,' Mr Taylor said. 'I'm sure Miss Hutchins's mother parted with her for the same reason.'

'Well. Parents must make sacrifices, I suppose.' Louise sounded almost meek. She likes him, Ellen thought.

'Where did you go to school, Mr Taylor?' she asked.

'The French school, in Cork. Our family lands are not far from there, in Kerry.'

'I also went to a French school,' Ellen said. 'Madame Praval's, at Platanus on the Donnybrook Road.'

'The Donnybrook Road?' He raised his eyebrows. 'That's in the country.'

'Yes, quite isolated. The nearest building is a nunnery, across the fields.'

'A curious mix, nuns and schoolgirls.'

'We never saw the nuns. Nor, I'm sure, did they see us.'

'I've heard of gentlemen pulled from their horses on that road, robbed and beaten, left for dead,' Louise said.

'We were cautioned never to open the front windows,' Ellen said. 'Or venture into the garden after dark.'

'Fear is an effective method of containment,' Mr Taylor said.

'What can you mean, sir?' Louise asked. 'Surely it was for the young ladies' protection.'

He shrugged, caught Ellen's eye, smiled, then sat forward and said in French, 'Or were they fearful what you might do if you had liberty?'

'What are you saying, Mr Taylor?' Louise asked. 'My French is poor.' Her voice was amused, but Ellen knew her well enough by now to detect a peevish note.

'Excuse me, ma'am. I simply asked Miss Hutchins if she'd enjoyed her schooling. I wish to gauge her competency in the language.' His eyes widened. '*Alors*, Miss Hutchins?'

Conscious of Louise's stare, she replied, also in French, 'We were too hungry and frightened to go far, even if the gates had been wide open.'

He nodded, and she caught something in his expression – sympathy? surprise? For he had surely thought to shock her. Turning to Louise, he said smoothly in English, 'I exalted in my schooldays, Mrs Stokes. No doubt there were some similarities with Miss Hutchins's, but also many differences. Young gentlemen, for example, spend more time outdoors.'

Truly, Ellen thought, he had nerve. She felt suddenly curiously enlivened. 'Doing what?' she asked.

'What they've always done. Riding, hunting.'

'You enjoy riding and hunting?'

'Not particularly, as it happens. I prefer more scholarly pursuits.'

Louise smiled thinly, obviously bored with the line of conversation. 'May I ask what age you are, Mr Taylor?'

'Eighteen.'

'Ah. The same as Miss Hutchins. My husband thought closer to twenty.'

'I'm often taken for older than I am.'

At that moment the door opened and Dr Stokes walked in. He came across the room, extending his hand.

'You're welcome, young Taylor,' he said. 'Excuse my not being here when you arrived. My appointments ran later than I intended, due to unforeseen complications.' He went to the sideboard, poured two glasses of wine.

'You need not apologise, sir. I too will have to get used to managing such "unforeseen circumstances", for I intend going into medicine myself.'

Dr Stokes handed him a glass. 'It's a rewarding profession, though not ideally suited to family life, as my wife will attest.'

'Dr Stokes gives so much of himself unstintingly to the citizens of Dublin, I would not dream of complaining,' Louise said.

'Have you been reminiscing with your cousin?' Dr Stokes asked Ellen.

'In French, no less,' Louise said.

'That means you're already good friends,' Dr Stokes said. 'French is the language of intimacy. Not like English, seemingly designed for preventing true expression.'

'The common people express themselves quite freely,' Mr Taylor said. 'Have you noticed, Miss Hutchins?'

'Servants, you mean? They generally say very little, unless they are amongst themselves.'

'Many of our countrymen have a talent for disarming one with their candour, even while being most duplicitous,' Louise said. 'I prefer blunt speech once the sentiment is genuine.'

'As with so many things,' Dr Stokes said, 'Truth lies somewhere in the middle of all opinions.'

A soft knock on the door, and the manservant, Plunkett, entered. 'Dinner is served, sir,' he said.

'Come, young Taylor,' Dr Stokes said. 'We keep a well-stocked table, much of it edible.'

The first course had been laid out – roast beef, spinach and onions, biscuits and pickles. Plunkett went around with the soup. 'Have you always wanted to be a physician?' Ellen asked Mr Taylor.

'It's really my father's ambition. I'm agreeable to it. It's an advancing area of science. It will take many years of study. First I must get my B.A., then I'll specialise. Though I confess it's not my main area of interest.'

'No?' She sipped at the lukewarm soup. Flecks of fat and thyme floated on the surface.

'While my fellow students were thrashing through fields on horseback behind a pack of dogs, I spent my time collecting and identifying plants.'

'Botany,' she said.

'Yes. Botany.'

'You've come to the right house,' Louise said.

'Compared to Dr Stokes, I consider myself an amateur.'

'You're well on your way, Taylor,' Dr Stokes said. 'I've been recommending books on the subject to your cousin, in the hopes I might convert her also.'

Mr Taylor set down his spoon. 'Which volumes?'

'Rousseau's *Letters on the Elements of Botany, Addressed to a Lady*; Mrs Wakefield's *An Introduction to Botany*.'

'I'm not familiar with Mrs Wakefield.'

'Her writings are intended for ladies,' Dr Stokes said. 'I read through a few chapters. Practical, sensible, if overdone, in my opinion.'

'How so?'

He glanced at Ellen. 'Overly cautious. Women aren't children.'

'No, indeed,' Mr Taylor said. 'And what is your opinion of Rousseau, Miss Hutchins?'

She wiped her lips. 'The parts about nomenclature are confusing. I would scarce have thought that a plant could contain so many elements, all precise, capable of deviation in so many aspects.'

'You find it dull?' asked Mr Taylor. He helped himself to a slice of beef.

'Not at all,' she said. 'The forces of creation, represented in the smallest of living things, each minute part isolated and named, described and annotated, set down as thus, its place in the greater scheme, to be understood as part of the whole ... I can see it's of great fascination.'

He nodded. 'Bravo.'

'Miss Hutchins,' Louise said. 'That's the longest speech you've made since you came to stay with us.'

'So,' Mr Taylor said. 'Has Dr Stokes tempted you to join us?'

She looked at her plate. 'I have not enough learning for that.'

'You have education enough. Besides, botany should be studied beyond the library or chamber. You must get out into the fields and woods, you must match the theoretical with the physical.'

'Mrs Wakefield would agree with you. She writes that, *books should not be depended upon alone, recourse must be had to the natural specimens growing in fields and gardens.*' Though, she thought, I can't wander where I please, as you do, Mr Taylor.

'On that point at least, Mrs Wakefield speaks sense,' said Dr Stokes, 'Though it's not always easy, in this climate. Oh, to be Rousseau, in the mountains of Europe, every year a mild spring and dry summer!'

Plunkett carried in the second course: pigeon, a bowl of stewed plums and cream. Louise had made an effort. She dipped her head, and Plunkett came to take the glasses for refilling. Ellen had not touched hers, and indicated he might take it away. 'Tell us more about your childhood in India, Mr Taylor,' she said. 'How exotic it must have been.'

'I have only impressions, vivid but fleeting – immense heat, lying in a cocoon of muslin designed to keep out mosquitoes, with an ayah crooning some native lullaby. Keeping to the shade to avoid the brutal sun.' He looked at his hands. 'My brothers and I killing a snake we found curled asleep on the veranda, with a rock.'

'Oh.' Ellen made a soft sound.

'I know,' he said. Ellen watched his eyes shift from deep brown to green. 'Even though it was probably poisonous, we were sorry for it after. I find it disagreeable to kill an animal to this day.'

'How cruel boys are.'

'Still,' he said, 'afterwards we were given a mongoose as a pet, a belligerent creature resembling a stoat.'

'You must have been lonely when you first came to Ireland,' she said.

'I thought I'd die of it,' he said. 'But I have the Atlantic, as well as India, in my blood. I'll settle on that particular coast one day, and build a house.'

Later, the ladies left the men to their talk. Louise took Ellen's arm and drew her aside. 'You may feel obliged to humour Dr Stokes by displaying an interest in botany and such like,' she said. 'But take care not to place yourself in that tiresome category of young woman.'

'Which is?'

'The self-improver, the would-be intellectual. At your age, it's unappealing, and risks appearing immodest.'

'I'm very grateful to Dr Stokes,' Ellen said.

'Of course. He converts wherever he can, that's his nature. But another kind of man may admire a clever woman and yet dismiss her.'

'Dismiss her?'

'As a possibility.'

'I don't understand.'

'Oh, Miss Hutchins, you will.' She squeezed Ellen's arm. Ellen looked down at her hand, taken aback by the uncharacteristic gesture of warmth. 'Think on it, my dear. I, as well my husband, have your interests at heart.'

Dr Stokes sat at the desk in his shirtsleeves, transcribing notes. As he scribbled, he said, 'What did you think of Mr Taylor?'

Louise yawned from the bed. 'Pleasing, I agree,' she said.

'You seemed quite charmed.'

'I wonder what she made of him. She is so ... inexpressive.' After a pause, 'Did you not say that her mother and father were also cousins?'

He put the quill down, turned around. 'Let her enjoy his company without complication. Mr Taylor will be a student for many years. They're young. Too young.'

She sat propped against the pillow, her hair about her shoulders. 'I don't know what you mean.' Her eyes widened, guileless. 'All I suggest is that he might chaperone her on occasion. She can't spend all her time in your library. As her cousin it would be perfectly proper. Don't you think?'

He rubbed at his eyes. He'd risen at six o'clock and spent most of the day in the dispensary in Temple Bar, attending on the poor and diseased, for most of whom he had little hope of healing their poxes, coughs and stuttering hearts. Resolving the effects of poor diet and filthy air lay beyond his powers. 'I have no objection,' he said. 'If she finds him agreeable.'

THREE

Just before Christmas, a letter arrived from Caroline. She'd left school, briefly for Henrietta Street, then to Bury Hall, her father's estate in Kildare. She begged Ellen to visit, *though it's too cold to travel now. Spring, come with the spring.*

Spring seemed an impossible dream. On Christmas morning the temperature hung well below freezing and barely rose throughout the day. Janey said the water issuing from the pump in the yard had reduced to a trickle. Throughout the church service, Ellen's thoughts slipped away from her prayers; the effects of the early hour – yawning could be heard throughout the church – and the intense cold made it impossible to concentrate. There would be a special dinner later at Number 16. She had been given a new dress: blue cotton that brought out the colour of her eyes, or so Louise said. A parcel had arrived from Cork, containing dried herring, a fruitcake, hand-stitched handkerchiefs.

Mr Taylor promised to take her walking when the weather improved.

She forced herself to concentrate on the sermon. The infant Jesus, born for the salvation of mankind. Brotherly love, forgiveness: a soul clean as snow, fierce as fire.

During the short journey home in the carriage, she huddled together with the children. Even buried under swathes of shawls and blankets, their teeth chattered. The shops along Grafton Street remained shuttered, though hardy souls scuttled along in clouds of their own breath. The carriage trundled past the dilapidated stone walls of St. Stephen's Green. Too cold to walk today, the ponds frozen over, snipe and ducks sheltering in the shrubbery.

Back home, Janey brought a heated brick to Ellen's room. 'For your feet, miss. At least I get to thaw out now and then beside the fire.' She wrapped the brick in a blanket, placed it at Ellen's feet. After an hour, Ellen could freely wriggle her toes.

Before dinner the servants assembled on the stairs, lined up according to rank, Plunkett straight as a pillar at the head of the row. Dr Stokes thanked them for their labour throughout the year and presented each with a coin. Then they dispersed again, back to the kitchen and scullery, their stations in the hallway and dining room.

Louise allowed the children sit at the Christmas table, with the various guests – colleagues of Dr Stokes, an elderly cousin, bachelor acquaintances. The candlelight caught the soft shine of the children's hair, their bright eyes. Minnie overcame her awe and stretched her little arm across to snatch a cherry and place it whole in her mouth before anyone could reprimand her. As plates of goose were passed around, Ellen's eyes strayed to the empty place setting beside her. Quickly, she fretted through the possibilities: an accident, a student scrape gone wrong. He bled by the side of the road, crushed under the hooves of a runaway horse. He lay unconscious in one of the chaotic, filthy city hospitals. He'd succumbed to a fever in his sleep.

Worse: he'd forgotten. Or accepted another invitation.

Still, in the midst of the chatter, clinking glass and scraping of forks on china, she didn't hear the door opening, so that when she looked up and saw him standing there, she burst out, 'Mr Taylor!'

'Here you are, already assembled. I am late.' He seemed to bring a cold vigour into the room, rubbing his hands together, moving about the table to shake hands with Dr Stokes.

His nose and ears were pinched red. 'Can you forgive me, Mrs Stokes?'

'You don't sound penitent, Mr Taylor, only exceedingly cheerful, and for that you are welcome,' Louise said. 'For Miss Hutchins's sake, as well as our own.'

'I studied late, and bumped into an acquaintance on the way,' he said, slipping into his chair beside Ellen; leaning sideways he said, 'You look very well, cousin. Is that a new dress?'

'You look frozen, Mr Taylor.'

'It's like India in here, I'll warm up presently. Ah, goose!' He speared a fat leg and dragged it, dripping grease, onto his plate.

'You shouldn't stand about outside in this weather.'

'The night is glorious. The stars sparkle with great intensity. Under just such a sky must the wise men have journeyed to find the infant Christ.' He seemed almost euphoric, his eyes glittering.

'The wise men might not have been at risk of catching pneumonia.'

'Miss Hutchins.' He shook his head in mock sorrow and lifted the wine glass Plunkett had just filled. 'So practical, so serious. If I catch cold, I will only have myself to blame. I considered the enjoyment of the night sky worth the risk.'

'Would you not give me the same advice?'

'Perhaps, but I am to be a physician, so mine would carry more weight than yours. Besides, your dress is in no way suitable for night-time star gazing, whereas I have stout boots and a decent coat.'

'You are in high humour.'

'It's the sight of all this food. It makes me dizzy.'

When Plunkett carried in the trifle of jewel-coloured jelly topped with layers of cream, fruit and nuts, and set it on a silver stand, everyone, including Ellen, clapped. After midnight, as the guests left, shouting muffled farewells, she wrapped herself in her shawl and ventured as far as the bottom step to watch them go.

'See, Miss Hutchins. The sky has clouded over, blotting out the stars. And it begins to snow.' Mr Taylor held out his hands as if to receive a gift. She looked to the sky, back along the high front of the house. Clouds of flakes floated down as if from the roof itself. Her lashes filled with ice and she blinked downward, blind and laughing. 'So you're not always serious,' he said at her ear. 'You should laugh more often.' Despite the freezing air, and much to her irritation, she felt her cheeks grow hot. He had this ability, to charm and provoke at the same time. She would not, for any inducement, have admitted to him that she liked both, in equal measure.

❧

If she had been craving the open outdoors, the Phoenix Park proved enough to satisfy. 'It's as though we were in the country,' she said, peering out the carriage window as they rolled along the central avenue, past mature sycamore and horse chestnut trees, still winter-bare. The park boundaries lay out of sight on all sides, far in the distance.

'Over seven hundred hectares,' Mr Taylor said. 'Shall we leave the carriage and walk?'

They strolled for some minutes along the path, obliged to nod and smile whenever they met other walkers. 'You remembered to wear stout shoes?' Mr Taylor asked.

She lifted her skirts to show him. An old pair belonging to Mrs Stokes; too small, they pinched, but were better than any Ellen possessed.

He frowned. 'They'll have to do. Now, if you are willing let's get away from all these promenaders.'

'Gladly.'

They veered off the path. Away from the avenue stretched open meadowland, patched with copses of young trees. Winter, though lingering still, had undeniably begun to melt away; a thin skein of green brightened the hedgerows, the grass gleamed fresh. Mr Taylor set a brisk pace and Ellen quickly grew warm.

'Early yet to collect any plant samples,' Mr Taylor said. 'Though I'm always looking. The botanist's affliction.'

'How did you begin?' Ellen said.

'We covered the basics in school,' he said, 'where I was fortunate to have access to a microscope. Then it was a matter of persevering, with whatever books I could lay my hands on, and the meeting of like-minded teachers like Dr Stokes. Come, if we go this way, we eventually enter woodland.'

The branches grew dense and merged overhead, blocking the light. A deepening silence, like walking into church. She glanced about; they were now quite alone. With the change in atmosphere, she felt suddenly conscious of every small sound, his breath, coming faster as they picked up speed, the crackle and snap of twigs underfoot. Rustling in the undergrowth, a cacophony of squawking, twittering, cawing and flapping; birds becoming restless, anticipating their spring nesting, making the most of the daylight.

'As I told you, I consider myself a beginner,' he said suddenly, breaking the spell. 'Though already I'm considering making the bryophytes – mosses, hornworts, liverworts – my speciality. Fortunately our damp climate offers many opportunities for study.'

She nodded, conscious of the wet spreading through the soles of her shoes. Under the shade of the trees, the ground had turned to sludge. She picked her way along, avoiding the worst of it. Mr Taylor trudged on, his boots squelching.

Louise's warning flashed into her mind. If she didn't speak at all, she risked looking like an empty-headed fool. She had no gossip, no amusing stories. She liked to read but didn't trust her taste; she knew nothing of politics beyond common trivialities. She had never been anywhere, knew nothing of the world.

He will think ill of me or not, she thought. She said, 'I have read of these plants, in Rousseau. The class of *cryptogamia* – they have no flowers whose parts are visible to the naked eye?'

'That is precisely what intrigues me.' He swiped at a bush with his cane. A small brown bird flapped away, chattering alarm. 'From the Latin *crypto* – hidden, secret; they have a mystery that flowering plants do not. Their fructification is concealed, they hide their treasures from all but the most determined student.'

'Of which you are one?'

'During the summer months, when I head out into the woods, fields and mountains. This is the point: you cannot hope to study botany from the pages of a book, even books as fine as Dr Stokes holds in his library. The science will prove too remote. You must experience the riches of God's creation with mud on your boots, in order to make any sense of it.'

In that case, she thought, I am well on my way. To her dismay, her skirts had trailed into a puddle. Her lower legs grew numb; she suppressed a shiver. 'Rousseau says that the field of *cryptogamia* is particularly difficult,' she said, 'That plants of that class must be searched for in places, and at a season, that might prove harmful to one's health. He recommends more gentle exercise with regard to the study of nature.'

'He addresses himself in his fictional letters to a lady, and, as we know, ladies are not so robust as men. Though I should think such advice would only serve to arouse curiosity, rather than dampening it. We're never so inclined to venture somewhere as when we are implored not to for reasons of good sense. At least, that is how I am. Ah!' He stopped abruptly and leaned over to examine a tree trunk, of no obvious interest. '*Tortula papillosa*. I can loosen it … like so … yes. I have it.' He turned, smiling broadly, and showed her a pincushion of greenish moss on his knife blade. With his free hand he groped in his pocket. 'Blast,' he said, 'I have come out without my handkerchief.'

'Take mine,' she said, pulling it from her sleeve.

'I'll return it.'

'No matter, I assure you.'

He spread it out on his palm, loosely wrapped the moss. 'That will do until I get back to my rooms.'

'Is it rare?'

'Not particularly. It's a fine specimen, though.'

'It seems to please you greatly.' His cheeks were flushed, his expression that of contented satisfaction, such as little Charles had when occupied by a favourite game. 'I can see it might make a walk more diverting.'

'Walking with my cousin is diversion enough,' he said.

'Fibber. I doubt you ever have "diversion enough".' He laughed, and glanced at her sideways. *See*, she thought, *I have your measure.*

They walked on for what must have been an hour, Mr Taylor stopping regularly to poke and examine at plants, though without finding anything else to inflame his interest. Finally they emerged from the wood into the light. 'We have almost come full circle,' he said. 'Soon we'll come back to the carriage.'

'I had no idea the park was so vast.'

'How many years in Dublin and you've never been in the Phoenix Park?'

'I have scarcely been anywhere,' she said. 'We were constantly told how dangerous the city is. Gangs of men fighting in the streets, pickpockets, drunkards ... '

'Aye, well, there's all that, at times. Trouble flares up and dies down again. The city simmers in a state of anxiety.' Ellen looked back where they'd come from, the thickets of grey tree trunks merging into shadow. Mr Taylor cleared his throat. 'But a lot of these tales are overblown, for the sake of selling newspapers. Take my arm, cousin.'

They walked on. 'To have seen so little of the place you live in must be disorienting,' he continued. 'A sense of place is crucial, I think. It gives us a sense of what is relative, allows us to think beyond our own limited existence. Did you not crave ... expansion?'

'Probably more than I realised.'

'You're like a prisoner, released.'

'Now you're exaggerating.'

'Only slightly. Why did you not ask to go beyond the walls of that school?'

Surely he knew most young women were, in one way or another, confined? For all his confidence, at times Mr Taylor had the air of a knowing but sheltered schoolboy.

Back in the carriage, catching a glimpse of her shoes, he exclaimed. 'How wet are they?'

'Not very.'

He bent suddenly and grasped for her foot. She gasped, tried to draw it away, but too late; he had seized it in both his hands and pressed upon it. 'Mr Taylor, what are you doing?'

He ignored this. 'Completely soaked through,' he said, frowning. 'Why did you not say anything?'

He released her foot, she drew it safely back under her skirts. 'I was enjoying our walk,' she said, stiffly.

'Let's hope you don't catch cold. Mrs Stokes will hold me responsible.' He seemed quite unperturbed – was he unaware of his impertinence?

'If I catch cold it will be no one's fault but my own.' She hoped he detected the annoyance in her voice. 'As you said yourself on Christmas night.'

'Ah. I knew you had not forgiven me for that.'

'Well, we are now even, Mr Taylor.'

'Could you not call me Thomas?' She didn't answer. 'Tom, even, as my friends do. We are cousins. I think it would be proper.'

'We haven't known each other long.'

'Long enough. And I shall call you Ellen.'

He was incorrigible. Rude one minute, winning the next. To refuse seemed impossible. 'Very well.'

'Tom.'

She suppressed a smile, kept her tone serious. 'Very well, Tom.'

'Excellent, Ellen.' He settled back against the seat opposite. 'Tell me more about this school of yours.'

'I doubt you'd find it amusing.'

'The people who ran it, then. A Madame Praval, you said?'

She raised her eyebrows. 'Well remembered.'

'The name registered. I've been told of a Frenchman, Charles Praval, who taught in Dublin.'

'Madame's husband, long dead.'

Tom leaned across, one arm on his knee, eyes wide with interest. 'I've heard that Charles Praval sailed with Sir Joseph Banks on the *Endeavour* voyage.'

'Madame sometimes mentioned her husband's acquaintance with Sir Joseph Banks. A curious print hung in the front hall at Platanus, of a sort of primitive island camp. Monsieur was supposed to have drawn it on the voyage.'

Tom's mouth hung open. 'Did you not encourage her to tell more?' he said.

'She was a peculiar woman. Forbidding, though not unkind. She wore black for many years after Monsieur died, long after the mourning period. We didn't speak of him, unless she did. Of course, now I realise the significance of Sir Joseph Banks. But then …'

'Then?'

'It seemed so improbable, so far from Dublin.'

'Men have often led more interesting lives than their present circumstances might suggest.'

'So it appears.'

'Were you fond of her? Madame Praval?'

Ellen looked at her hands. 'I felt gratitude. Fear. Fascination. Anything more … she was too remote.'

'And your present guardians? How do you like the Stokeses?'

'They've been very kind.'

'You wouldn't like to go home to Cork?'

'My brother Emanuel wants me to stay here for the present.'

'But if you could choose?'

She hesitated. To say she had begun to consider 16 Harcourt Street home seemed disloyal. 'For the present, I'm content.'

'Your brother and Dr Stokes are old friends, are they not?'

'Yes. They met in Trinity. Dr Stokes was a kind of mentor for Emanuel.'

Tom raised his eyebrows.

'Why do you look at me like that?' she asked.

'You don't know the circumstances of their friendship?'

'Dr Stokes was a lecturer, Emanuel a student.'

He leaned forward. 'You must know that Dr Stokes was an early member of the United Irishmen.'

Ellen said nothing.

'You've heard of the United Irishmen. They instigated the '98 rebellion.'

'I know that much. But you must be mistaken about Dr Stokes. He never discusses politics that I hear. He appears in every way a model citizen.'

'He is, now. And, of course, he's careful with whom he speaks. He was a moderate, apparently, never one of the main players. But for a long time afterwards he was smeared with suspicion. He lost his teaching position in Trinity for three years.'

'Who told you this?'

'It's common knowledge, I assure you. Dr Stokes told me some of it himself.'

'And Emanuel?'

'That's less clear. What I do know is that he was an aco-lyte of Wolfe Tone's.'

'The revolutionary?' He nodded. She shook her head. 'You have heard wrong, Tom.'

'It's true.'

She looked out the window. Her mind raced. After a moment she said, 'Wolfe Tone died, in custody. By his own hand.'

'So they say.' After another short silence, he said, 'I've upset you.'

She shook her head, forced a smile. 'No.'

He struck his fist on his knee. 'Thoughtless of me. He is your brother.'

'I can't believe he was involved in any of that.'

'What do you remember of that time? You were only …'

'Thirteen.' Whispers, drawn faces, candlelight flaring in the small hours of the morning. Fear permeating the school walls and leaking down the chimneys. 'We were told very lit-tle. I heard the servants talking, once or twice.'

'We were safe enough,' Tom said. 'Boys being boys, we were rather excited by it all.'

'Excited!'

'Many young men thrill to the idea of violence, before they've actually experienced it. Factions developed on either side, argu-ments that came to blows. Reflecting the whole country.'

They had turned onto College Green, passing Trinity College. 'Almost home,' she said.

He bent forward. 'Before we part, Ellen. What I said shouldn't alter your impression of Dr Stokes. He's a man of impeccable character and moral strength. Look, for example, at his work in the dispensary. His past actions stemmed, I'm sure, from a genuine love of Ireland. As well as a sense of outrage at the unfair treatment of Catholics, rather than any thirst for anarchy.'

'I never imagined Dr Stokes capable of passion for aught but botany.'

He smiled. 'Ah, but we've just agreed that any man may have more to him than first appears.'

'A thought I will consider from here on,' she said. After a pause, 'And you? Have you similar leanings?'

'I've learned from the experience of others. I abstain from politics. However much I may question the system.'

'I'm glad to hear it. You will have more to offer as a physician.'

'Yet if there is ever to be change, someone must stick their head over the parapet.'

'It's natural to prefer a peaceful existence rather than seeking out turmoil.'

'Until turmoil forces itself on us, at any rate.'

At Harcourt Street he saw her inside the house but would not stay. 'I am meeting friends, Mrs Stokes. I fear I have worn out your ward with fresh air.'

'Call again, Mr Taylor, whenever you like.'

The next day a packet arrived for Miss Hutchins. When she opened it, she found her handkerchief, laundered and neatly folded. He had included a note: *I said too much. My confidences were well meant if overly explicit. I hope they did not spoil an enjoyable afternoon in the park. Tom Taylor.*

Louise became ill. She fled from the table, white-faced, with her hand over her mouth. Some mornings she didn't appear at all. Her eyes grew dull, her energy shrivelled. When she did leave her room, she lay against the sofa cushions in the drawing room, dozing or looking out of the window. Dr Stokes asked Ellen to help with the children, to divert them from their mother. She did so, gladly. Soon after Louise confided that she was expecting another child.

'A blessing, wrapped in torment,' she said. 'I've been ill throughout all my pregnancies and it seems I shall be so again with this one.'

'I will help you, if I can.'

'Dear Miss Hutchins.' She laid her hand on Ellen's. She seemed softer, as if her pin-sharp edge had worn away. Their conversation after Tom first came to dinner, mystifying as it had been, seemed to have marked some kind of alteration in her regard for Ellen. Her manner had become that of an older sister, still distant but fond; she increasingly depended on her for help with the older children. And exhaustion, apart from all else, made her kind. Occasionally faint smears of vomit crusted the corners of her mouth. Her hair stuck in damp strands to her neck. She complained of feeling hot and bloated, that her shoes no longer fit, though so far she looked much the same as always, as Ellen assured her. She sighed. 'In time my body will become accustomed to its invader, and begin to expand.'

Despite all – the rushing from table, the queer smells, the complaining – she moved as if carrying a tray of fine china, as though in a state of grace. She rested her hands on her stomach, a drowsy smile on her lips, only half listening to what anyone said.

The air grew milder, promising a damp and gentle spring. Louise's sickness eased enough for her to get up with the rest of the household in the mornings. Her eyes brightened, her tongue sharpened once more. Caroline wrote again, entreating Ellen to visit. Now Ellen felt torn, half wanting to see Caroline, half reluctant to leave. She had established herself in the household, they needed her, even with Louise so much improved. And how pleasant to be needed! Yet underneath this legitimate, selfless reasoning lurked something less tangible, a truth she could barely acknowledge even to herself, as a flitting shadow at the corner of her vision, that undeniably coloured her feeling: the prospect of the deprivation – and deprivation it would be – of Tom's company. The new term progressed, however; a diligent student, he became absorbed in his books, university life, he had many friends and demands on his time. Finally a week came where he didn't appear at all at 16 Harcourt Street. She told herself to be grateful for the attention he had showed her, and wrote to Caroline, *Expect me the second week of March.*

❦

A twenty-three-mile journey on a well-maintained turnpike road: when the coach stopped at Naas, Caroline and her father were waiting in their gig. Caroline had filled out, her round face framed by a silk bonnet. She seemed excited and nervous, chattering like a bird the entire way along narrow, meandering country lanes. Ellen nodded and smiled, silently naming the flowers in the hedgerows – celandine, violet, cuckoo pint, primrose, speedwell. The knife-clean smell of new grass, the light sharp between the rushing clouds. Caroline's voice, gay and desperate, was a tide upon her senses. Sleepy-looking cows raised their heavy heads to stare at the carriage as it passed.

At last, the black iron gates of the Bury estate. Thick trees cleared to reveal a large house of grey stone: handsome, austere, hard edges and symmetry, many windows. A servant came running to take the horses. When they'd climbed down from the gig, Mr Bury disappeared towards the yards leaving them alone. Caroline led Ellen into a cavernous front hall of dark flagstones, decorated with relics of hunting, fighting and farming: deer heads, antlers, hides and horns, guns, pikes and swords; portraits mainly of horses and dogs, few of people.

Caroline whispered, 'Ugly, isn't it? You may say so, you know. Father won't pay to redecorate. He begrudges the expense.'

'It reminds me of what I remember of Ballylickey, though this house is more grand.'

'I'm bored to death of it. We didn't even go to town for winter, like everyone else did. You are fortunate.'

'All I do is amuse the children, read, draw, take the air once a day.'

'At least you have congenial company. The local gentry wipe their mouths on the tablecloth and smell of mould and spaniels. When I hear a carriage on the drive I don't know whether to hide upstairs or run to greet it.'

In the great echoing rooms, their whispers seemed shockingly loud. Flat countryside lay beyond the windows, nothing but trees to break the horizon.

'I'm glad you've come,' Caroline said, squeezing her arm.

'As am I,' Ellen said, rousing herself. 'I look forward to long, cheerful walks in the grounds.'

'Then you must be prepared to get mud on your skirts.'

'That I don't mind.'

In the gloomy drawing room, a dog rose from in front of the fire and trotted across to sniff under Ellen's skirts. Caroline snapped her fingers. 'Get away, Charming, you great oaf.'

'What a name!'

'His looks are plain – in that he matches the house – but he is harmless.'

Ellen held out her hand. 'Let's be friends, Charming.'

'You look better than when I last saw you,' Caroline said. 'You have good colour.'

'I feel well.'

'I wish I'd been taken in by a kind city gentleman. Father scarcely notices me. If I neighed or barked, he might pay me more attention.'

'You don't mean it.'

'I do.' Her eyes grew serious. 'Do you still get headaches, or strong pain?'

'Almost never,' Ellen lied.

'Hopefully you've grown out of it, as we often do of our childhood ailments.'

At dinner Mr Bury, a giant of a man with a high complex-ion, chewed slowly and loudly on his lamb chop and stared directly ahead. His knife and fork almost disappeared in his great fists. There had been no conversation, only the clang of cutlery on china and the clunk of his wineglass set down on the tablecloth. The meat was excellent, tender and lean, but so oppressive was the silence Ellen could hardly taste it. Summoning her courage, she swallowed and said, 'Your house has a fine situation, sir.'

He nodded in acknowledgement.

'The trees are beautiful,' she went on. 'How peaceful the sheep in the fields look.' Caroline rolled her eyes. 'And you have wonderful walks at your disposal.'

'I'd swap all the sheep in Kildare for a pleasant city street,' Caroline said. She drew her fork across her plate with a squeal. 'From school to this, one prison to another ...'

'Enough, Miss!' Mr Bury's voice was no less strident for his mouth being full of potato. Caroline went white, then red, and clutched at her napkin. Ellen tried to catch her eye but she stared down at her plate.

When she'd been a child, her father had been so tender and attentive. What had changed?

After dinner, as they sat in the drawing room before a smoky fire, a gaggle of neighbours called; two elderly country gentle-men, their wives and a stout middle-aged woman, daughter to one of the couples. They gawped at Ellen as if she were a new species, then drew up chairs around the fire, artfully forcing her to the back of the circle. The talk comprised entirely of the rapid-fire exchange of names and gossip, so coded Ellen couldn't follow. Everyone appeared to be related. Occasionally the speaker detoured to expand on a finer point, looped to another detail – the lineage of a man or horse, the inhabitants of such a house before such-and-such, the reason so-and-so no longer spoke to so-and-so, local assaults, murders – until the thread of the original tale was all but abandoned. This didn't seem to bother anyone. To Ellen's surprise, Caroline had perked up. Perched at the centre of the gathering on a yellow cushion, she became animated, waving her hands and tossing her curls. Ellen followed the weave of conversation for a time, before faltering and getting lost. She allowed the voices to drift around her.

A flabby-faced man with fair whiskers had arrived later and now stood by the wall in a dark corner, leaning against a dresser. He grunted when spoken to – 'What say you, Aylmer?' – Mr Bury crossed the room several times to confer with him in a low voice. Side by side, they looked like brothers, though Mr Aylmer's girth had not yet turned to fat.

Voices rose and fell like the rattle of nesting starlings. Ellen glanced at the clock: one o'clock. When would they leave and allow the household to get to bed? Some time later – minutes, hours – she heard Caroline say, 'My friend from Dublin is not used to sitting up late,' and jerked her head up from where it had slumped onto her chest. They swivelled in their chairs, startled, glaring; they had forgotten her. The room was cold, the candles reduced to sputtering nubs. Grudgingly, with much muttering, the visitors stirred, stretched, called for carriages. Charming trailed them to the door, as if to see them off.

On the way to bed Ellen asked about Mr Aylmer.

'He's a friend of Father's,' Caroline said. 'Quite old, in the late thirties. Though he has a fine house near Prosperous, he seems fonder of our fireside.'

'Your Father seems to esteem him.'

'Naturally – he has a thousand acres. What he lacks is wit and conversation.' She checked herself. 'I sound harsh. He's clumsy, that's all. Harmless.'

'Like Charming?' If Caroline stepped on Mr Aylmer, as she frequently did the dog as he slumbered around the house, would the gentleman lie there, just as placid and uncomplaining?

The rain held off for a morning and Ellen decided to sketch the front view of the house; Louise would be curious to see it. Servants set chairs out on the lawn and assembled a makeshift awning of sheets. While she sketched, Caroline played with a white, rather grubby, silk fan.

'Here's how I communicate to a gentleman that I want to make his acquaintance.' She moved the fan to her left hand and held it in front of her face.

Ellen put down her pencil. 'Caroline. You don't.'

Caroline shook her head, put on a haughty expression. 'I may have, once or twice. Subtly.' She laughed. 'Too subtly, for the gentleman in question didn't seem to notice. He looked at me, confused, as though I'd lost my wits.'

'I wonder where you learn such nonsense.'

'On the rare occasion that I'm brought anywhere, I must amuse myself.' She fanned herself, lifting the curls around her forehead. 'I wonder if Mrs Stokes tried such ruses on Dr Stokes, when they first met.'

'Unthinkable.'

'How many months gone is she?' Caroline asked.

Ellen frowned. 'I'm not sure.'

'So there will be yet another child in the house.'

'If all goes well.'

A thrush hopped across the lawn. Caroline called, waving the fan, 'Come here, pretty creature.' The bird regarded her with a beady eye and bounced away. 'Will it not get crowded?'

'The house has many rooms.'

'Wasn't your room the nursery?'

'At one time,' Ellen said.

Caroline: deceptively sharp, never as distracted as she appeared. 'Well,' she said, 'You're welcome at Bury Hall, for as long as you like. Forever, if I had my way.'

'I couldn't impose on your father.'

'He'd hardly notice if you took up permanent residency in the front hall. Besides, he likes you. He considers you my superior: "a modest, sensible young person". I'm vain and thoughtless, apparently.'

Ellen smiled. 'You're vain. But never thoughtless.'

'Not to you, dear Ellen. I'm careless with the feelings of others, those I love less. He chastised me for it.'

'To whom did he accuse you of being careless?'

Caroline shrugged, but didn't reply.

'He can't stay cross with you for long. Remember how he used to take you out from school when you were younger, and spoil you with sweets?'

Caroline hugged her arms about herself. 'That was then. Over the years he's become morose. He can be ...'

'What?'

'Oh.' She shook her head. 'No matter.' She assumed a pose, pouted. 'Vain as I am, when you've finished that, will you do a drawing of me?'

Ellen envied her ability to shake off gloomy thoughts. A spring sky, ever changing, light chasing darkness away.

One morning by her plate at breakfast she found a small package, tied in ribbon. She opened it to find a pair of fine leather gloves, pale blue with a stitching of flowers, and her name embroidered at the cuff.

Caroline laughed at her stunned expression. 'How could I forget, when your birthday falls on St. Patrick's Day? The

green the servants wear in their lapels will always remind me.'
Ellen pulled on a glove and flexed her hand, admiring the
snug fit, the soft gleam of the fine leather.

Later, Mr Aylmer visited; Ellen walked in to the drawing room and discovered him talking with Caroline. She
recognised the twist of irritation around her friend's mouth.
Seeing Ellen at the door, her face lightened. 'Ellen. Come,
please.' Mr Aylmer, interrupted mid-sentence, gave Ellen a
curt nod. 'Mr Aylmer has been telling me about his house
and grounds,' Caroline went on. 'He's having the lake stocked
with carp. I must say carp is not to my taste. I prefer trout,
or salmon.'

'I've never tasted carp,' Ellen said. 'I should like to try it.'

'I doubt you'd enjoy it,' Caroline said, wrinkling her nose.
'They spend their lives in mud and that's what they taste
of.' She changed the subject to the upcoming May Day, the
pole that the local people would erect and dance around, the
Queen they'd elect amongst the little girls of the village.

Mr Aylmer listened politely, though his shoulders had
slumped. Soon he bowed, saying he'd look for Caroline's
father, and left the room.

'Well, you succeeded in chasing him away,' Ellen said.

'Was I so obvious?'

'I've just realised I've not heard Mr Aylmer say more than
two words at a time before now.'

'Nor have I, to that extent. Before you came in he was telling me about the new plumbing he's installing, how much he's
spending on it. He was quite animated. I don't know what's
got into him. He's usually dumb as a statue.'

'You dislike him.'

'Not really. Perhaps. Oh, I don't know. I think he's a bore.'
She sighed. 'But what I think of him is beside the point. As
our neighbour, I should be cordial.'

'And he is a friend of your father's.'

'Yes.' She pulled on her lip. Then she threw back her head. 'Let's forget him and play cards. Or whatever you like, as it's your birthday.'

⁓

Movement by the door. A figure emerged from the darkness, dressed in white with long, loose hair, on tiptoe, like a dancer. Ellen sat up. 'Caroline?'

She padded across the room and climbed in Ellen's bed, making the bedsprings sigh, her body shuddering with cold. Eventually she became still.

'Are you ill?' Ellen whispered.

'I can't sleep.' Ellen had always envied Caroline her death-like slumber, impervious to thunder, other girls' nightmare shrieks, rattling windows. She turned over and faced her. 'Are you troubled by something?'

'I can't seem to … forget myself, like I used to. My mind boils with a thousand thoughts.'

'Such as?'

'Oh, different things. My mother, for one … I could not remember anything about her in school. Now that I'm here, I find memories returning, all the time, that I didn't know I had. Her face, her voice. She looks so pretty, so young … the strange thing is, she never lived here, but at Henrietta Street. Father built Bury Hall after she died.'

They lay in silence for a minute. 'What then,' Ellen said, 'can be the explanation?'

'I don't know. Except … she now wants to communicate with me, when before she did not.'

'The dead can't speak, Caroline. Much as we would want them to.'

'Can you be sure? Some of what she says is so vivid.'

'What sort of things?'

'Well … warnings. As if I were in a maze, and she cautions me against taking a certain path.'

'You are troubled. Will you not tell me the cause?'

'There's nothing. Except the fear I will disappear.'

'Disappear? What can you mean?'

Ellen would not have thought Caroline capable of such thoughts – that she'd admit to, at least. The counterpane rose and fell with her heavy breath.

'There are too many rooms in this house,' she said at last. 'Too many echoes. I fear I will be sucked away into the void, like a scrap of paper up a chimney. If that happens … how can I hold on?'

Ellen thought of the vast marble fireplace in the drawing room downstairs, down which the wind howled like some grief-stricken demon. She felt a prickle of fear along her spine, but made her voice calm. 'You're not making sense, Caroline.'

'Perhaps not. Yet it's how I feel.'

'Madame Praval advised me to try and be content, wherever I might find myself.'

'Oh, that old witch. Who cares what she thinks now?'

'Is it not sensible, though? Tell yourself: this is where I must be. Peace will come.'

'Will it?'

Ellen paused. 'Madame would say God fills that void you speak of. So that we don't have to.'

Caroline's voice faltered. 'I look for Him, too, in the dark. But I don't hear Him. I only hear my mother.'

'Perhaps it's the same thing.'

'Do you hear your father?'

'No.'

'Never?'

'Never.'

'Would you like to?'

'It wouldn't be natural.'

'But still ... your father. You might find comfort in it.' Ellen said nothing. 'Ellen, what do you hear, in the dark?'

'Naught but silence. Or my own inner voice, fretting in circles.'

'What have you to fret about now, safe in Dr Stokes's house?'

'The same reason you wake in the night. What will become of me. My place in the world.'

Her last day. Caroline followed her around the bedroom as she packed, pressing other gifts on her – ribbons, a note-book, Balbriggan stockings, silk and of best quality. 'Yours are threadbare,' she said, pushing them back into Ellen's trunk when she pulled them out, shaking her head.

'I don't like to ask Mrs Stokes. I'm already a burden.'

'You've said yourself, we have no choice but to do as we are told. Burden or no, is it our fault?'

'I suppose not.'

Caroline played with her hair before the glass, pinning it high, tilting her chin. Her sleeves pulled downward, showing her bare, white arms.

'What are those marks, Caroline? Did you injure yourself?'

Caroline half-turned from the mirror. 'What?'

'Those bruises.' Ellen pointed to the livid marks just below Caroline's elbow.

'Oh. That's nothing.' Caroline pulled her sleeves down, smiling. Deep puce had spread into her throat.

'They look painful.'

'Not very. I tripped in the garden and fell against a stone pedestal.'

'But you're hurt on both arms.'

'That's how foolish I am, falling twice in the same place.' She brushed past Ellen on her way to the door. 'I'll leave you in peace to finish packing. It's time. Much as I hate to see you leave.'

The Dublin coach had been directed to wait on Miss Hutchins.

While the driver loaded her trunk, Caroline pressed her wet cheek against hers. 'You never cry, Ellen, that I see. Aren't you sad?'

'You know I don't like to cry in front of others.' A habit learned in school. Held-back tears, swallowed sorrow.

'When will we see each other again?'

'Soon, I hope.'

They embraced once more, then Ellen pulled herself away and climbed on to the coach. She bid her fellow passengers good morning. They examined her with interest. She pretended not to notice and wriggled on the narrow seat, arranging herself against the hard wood and looking out the window.

Considering all, she was glad to leave Bury Hall. She discreetly wiped Caroline's tears from her face.

FOUR

William Stokes was born on the fourth of October.

Louise stayed in bed for weeks, while the baby lay swaddled in its crib beside her. The children were largely disinterested in the new arrival, though when Ellen brought them in to visit with their new brother, she saw Charles creep to the crib and yank at the tiny arm in a pretence of affection. The baby made no sound, though his eyes widened and his head jerked towards the sound of his siblings' voices.

Ellen arranged her room for studying in. Dr Stokes had a suitable table and chair placed by the window for best light. He proclaimed that she should have an oil lamp, as he did in his study, to save her eyes. She spent more and more time at her table, bent over books.

Being part of a family's happiness was even lonelier than being part of their sorrows.

One evening as she walked into the drawing room, Dr Stokes called: 'Miss Hutchins. I want to introduce you to James Townsend Mackay.' The man who turned to look at her had sparse fair hair and a sharp nose. She placed him at around thirty. 'Mr Mackay is Assistant Botanist at the Botanic Gardens of Trinity College.'

'How do you do, Miss Hutchins?'

'Mr Mackay has been touring our west coast, collecting rare plants,' Dr Stokes said. 'You don't know how envious I am of you, Mackay. How dearly I'd love to go on such an expedition!'

'You're welcome to join me. The countryside is exceptionally beautiful, if bleak and untamed.' He spoke with a pronounced, rolling accent.

'Alas, work and family concerns keep me here,' Dr Stokes said. 'The best I can manage are my weekend field trips.'

'Which have resulted in a fine herbarium, sir.'

'Where are you from, Mr Mackay?' Ellen asked.

'Kirkcaldy, Scotland.'

She had never met a Scotsman. 'What type of place is Kirkcaldy?'

'A ship-building town north of Edinburgh. The sea is the life-blood of the citizens.'

'Similar to Bantry, the town close to where I was born,' she said. 'A sleepy place cut off from the rest of the country. But open to the sea, and dependent on it.'

Mackay stuck his chin to one side. He had lively grey eyes. 'You visit it often?'

'No, sir. I have not been back there since I came to school in Dublin, many years ago.'

'A fantastic, wild place, Mackay,' Dr Stokes said. 'Though, as it happens, with certain unique climatic factors which result in a rare flora.'

'Perhaps I shall visit there some time in the future. The opportunities afforded by these unexplored parts of the country are wonderful, from a botanical point of view at least. It's precisely that which brought me to Connemara. I consider the trip to have been successful. Still, as many weeks as I devote to collecting, it's never enough, and therein lays the frustration. There's naught to compare with the native plant hunter. He alone has the time, over seasons, over years, to search out the most important finds, ever building on his local knowledge.'

'What is the purpose of your collecting, Mr Mackay?' Ellen asked.

'A fair question, Miss Hutchins. Without purpose, our endeavours are inclined to drift off course. Purpose is what keeps us going when disillusionment sets in. I intend on writing a book – the first modern flora of Ireland.'

Dr Stokes seized Mr Mackay's shoulder. 'An important enterprise, Mackay, and one that's overdue. You may count on me for any specimens I have from the county of Dublin.'

'I shall count on all my associates and friends across Ireland, to hunt on my behalf. It's a task beyond any single man. As with any catalogue, this book will be the culmination of the industry and perseverance of many.'

Walking into dinner, she heard Dr Stokes ask, 'How is that worthy and estimable gentleman, your master in the Botany Department, Professor Scott?'

Mackay shook his head. 'Not well. Though he doesn't complain, and does the work of two fit men, he coughs enough to break in two.'

'A pity so excellent a man must have his life's work compromised by a constant struggle with mortality. We must thank God for the privilege of our good health.'

'Indeed. You saw the dedication Dawson Turner wrote in his latest book on Irish mosses?'

'Turner, the English botanist? I have ordered *Muscologiae Hibernicae Spicilegium*, of course. He dedicated it to Scott?'

'Yes. Scott was much moved. Turner is a generous fellow – one who appreciates his friends.'

Dinner did not slow their discussion. Stuck down the opposite end of the table, Ellen tried to follow it. She could easily hear Mr Mackay, though he spoke in a moderate tone. Volume alone did not always mark out a voice as compelling.

⚜

She never received letters, so was astonished to find two beside her breakfast plate in the morning.

From Caroline: a cramped note announcing her engagement to Mr Aylmer. They would marry in the spring.

From Tom Taylor, on holiday in County Kerry: a brief enquiry as to her health, a report of his own (excellent), and an assurance of his regard for her at all times.

Louise was looking at her expectantly from across the table. Ellen folded Caroline's letter. 'My friend from school, Miss Caroline Bury, has become engaged.'

'Who to?'

'Mr Aylmer of Prosperous, Co. Kildare.' After a pause she added, 'A friend of her father's.'

'You know the gentleman?'

'I met him, yes. He is somewhat older.'

Something about her tone must have alerted Louise, for she merely nodded. 'And the other? Not bad news, I hope?'

'No, ma'am.'

Louise waited, then shrugged: no more information would be forthcoming. 'Well. Thank God for that.'

From his end of the table, Dr Stokes waved a letter he'd received in the same post. 'Your brother Emanuel, Miss Hutchins.' She started, sat straighter. 'Due back in Ireland for a visit within the month, though he cannot say when he might call.'

'You should like to see your brother?' Louise asked.

'Certainly, ma'am.' Feeling their eyes on her, she finished her breakfast, swallowing down the end of a bread roll with a mouthful of tea. It had lodged in her throat.

Poor Caroline. They had worn her down.

Tom Taylor had wasted an expensive sheet of paper and cost Dr Stokes seven pence for the postage, to say very little. Still, she hadn't felt like sharing even that with the table. Inadequate as they were, the scanty lines were hers, and hers alone.

As for Emanuel – he had not asked after her specifically, nor included a note for her. The prospect of a visit from any member of her family made her heart beat faster. From excitement, or trepidation?

She conceded: both.

Emanuel arrived closer to three weeks later. He sent no warning note ahead, and scarcely waited for Plunkett to announce him before striding in. A lean, alert, sharp-featured man in the mid-thirties with a full head of thick hair, immediately seeming to occupy more space, more air, than was merited by his height, which was only moderate. Ellen had been reading aloud to Louise from *The Vicar of Wakefield*; as he came in, she dropped the book on the floor. She fumbled to pick it up. Emanuel stood, waiting to be acknowledged, tapping his foot as if dissatisfied with the easy calm of the afternoon drawing room.

'Good day, Mrs Stokes. And is that my little sister Ellen scrabbling about on the carpet?'

She stood up, red-faced.

'Miss Hutchins, greet your brother,' Louise said, sharply. 'She's shy, Mr Hutchins, and probably overcome.'

'I don't bite,' he said. Though he made no move towards her, he held out his hand. Ellen walked forward and took it – would he want to kiss her? Should she kiss him? She lifted her face. They stood at nearly the same height. He shook her hand but drew back his head, studying her. 'I'd know my sister anywhere. Though she is now grown bonny and strong, I'm glad to say. She was a scrawny enough seedling.'

'You are much the same, Emanuel,' Ellen said.

He grunted. 'We are all older. Presumably wiser.' It was true he had more lines around the eyes and mouth, a speckle of grey in his light-brown whiskers. But he still had the intense gaze she remembered.

'I'm afraid you have missed my husband, he is out on his rounds,' Louise said.

'A pity. I am going now straight to Cork but thought to see him first.'

'And your sister, of course,' Louise said.

'Indeed.'

Louise indicated a chair. He sat heavily.

'How long will you stay in Ireland?'

'Some weeks. I have business to settle.' He nodded at Ellen. 'With our brother Arthur.' He was interrupted by noise from the street: wheels, the shouts of delivery men. Leaning back in his chair, he stretched out his legs.

'How do you find London?' Louise asked when the din had died down. Her tone was polite, stiff.

'I don't particularly care for it. Its sophistication is all surface. Too many people jostling for space and opportunity. Dublin is a village in comparison.'

'Still, how much more exciting for a man of your capabilities,' Louise said.

He looked at her, frowned. 'It's a cold place, madam, to make one's way. For now, though, I must stay. Samuel is still at school and needs supervision.'

'I dearly wish to see my little brother,' Ellen said.

'Little! He costs more than a good horse to keep. Our mother spoiled him. I admit, he shows promise, if he keeps his head out of the clouds. I keep an eye on him, when I can, though I'm too busy to play nursemaid.'

He gulped down three cups of tea when it was brought, but refused cake.

'Won't you tell us a little of London life?' Ellen said.

'I know few tales that a lady might find amusing,' he said, pulling at his collar. His eyes darted about. He clearly itched to go, though remained, held in his chair by an ingrained politeness. 'I cannot abide parties and avoid fashionable hoo-ha of any description.' He managed to identify enough mutual acquaintances with Louise to fill a half hour's conversation, though she continued to thin her lips and speak as though she begrudged every word.

At last the chitchat ran out. The silences grew longer. Ellen had many things she would ask him, had she the opportunity, and no Louise to watch, listen and judge.

'Well.' Emanuel stood, relief clear on his face. 'I must be on my way.'

'Dr Stokes will be sorry to have missed you,' Louise said.

He waved a hand. 'I will run into him somewhere in the city before I go back.'

She nodded. 'And what of Miss Hutchins?'

He paused. 'What of my sister?'

'Will you leave her here with us, or have you other plans for her?'

He folded his arms. 'You are content here, Ellen?'

She stared down at the rug, blushing. 'Yes,' she said.

'Then for the moment,' he said, 'I see no reason to change my sister's living arrangements.'

'Agreed,' Louise said.

After he left, she sat again, smoothing her skirts. 'Well, that was cursory enough. He seems much the same. At least he has the sense to leave you with us.' She glanced at Ellen. 'Perhaps you hoped he might take you with him to Cork?'

'No, ma'am.'

'Sensible. I believe your family's estate is horribly isolated.'

Ellen looked at her hands. 'I wouldn't want to cause them trouble.'

'Why should you?'

'Is not a dependent sister a drain on a family?'

'A girl may marry well, and raise a family's standing. You're young enough to have hopes of that. At any rate, you're better off here.'

Two weeks later a note arrived. It seemed that Emanuel had reconsidered his manners.

Ellen,

I will be in Dublin on Friday. I sail the next day. I want to speak with you before I leave on an important matter. Please arrange to be home at three o'clock.

Emanuel

At twenty minutes to three she stood at the window, behind the drapes. At half past, a loud knocking on the front door: she hurried to the sofa and arranged herself, book in lap. He came into the room less energetically than before, the lines of his face marked deeper.

'Stay sitting,' he said, as she put her book aside. 'I've had a trying journey back up. I'm not a man who demands easy living but the filth of some of these Irish hostelries is intolerable.'

'How goes it at Ballylickey?'

'Not well,' he said. 'Not well at all.' He did not sit, but paced about. In front of the mantel he stopped and picked up a marble paperweight, staring at it as if momentarily transfixed by its swirling reds and creams. He banged it down again. 'Jack can scarcely walk. He's been able to manage the house until now, with help from the servants, and Mother, to a certain extent. But he will be quite dependent soon.'

Ellen put her hand to her mouth. 'I didn't know he was so bad.'

'And Mother's own health is poor. She has become quite feeble.' He crossed the room and sat down opposite her. He had a scratch on one cheek, no doubt from the barber's razor. 'The situation is becoming untenable.'

'How terrible. Poor Jack.' Her thoughts raced. 'What does Arthur say? He lives close by, does he not?'

'They have concealed the worst of it from him. Besides, Arthur has established himself at his precious Ardnagashel with his family and appears utterly consumed with his own affairs.'

'Surely not to that extent.'

'You are ignorant as an egg, Ellen. You can't know what goes on down there. They are all quite odd, never having been anywhere or seen anything of the world.'

'I would help, if I could.'

'Well, now you have said it, I have a proposal.'

'Yes?'

'You shall go down there.'

She gaped at him. 'What do you mean?'

'What I say. To comfort Mother and help Jack. There's no one else.'

'You told Mrs Stokes that I should continue to live here, only two weeks ago.'

'I know what I said.' A flare of irritation, which he instantly dampened. 'There's no other solution. Initially I thought Mother and Jack could live with Arthur, but neither will leave Ballylickey. Mother has a deep attachment to the place. Sentimental memories of Father.' He jerked his shoulder in a shrug. 'Besides, as it happens, I don't want the house left vacant at this time. It suits me to have Jack continue as caretaker.'

'Why?'

'I have my reasons, none of which you need concern yourself with.'

She shook her head, confused. 'Is there truly no other way?'

'I'm afraid not.'

'If I wrote to Arthur …'

'It's my decision to make, Ellen, and I've done so. Had I known the state of things before I left for Cork I might have brought you with me. As it is, I'll have to ask Dr Stokes to make arrangements. He will oblige me, I'm sure. As he's always done.'

'How soon must I leave?'

'It can wait until the weather improves. No later than the spring, though. What's this? Tears?' He sounded exasperated, looked away as if repelled. 'Why must women always cry? It won't have the desired effect on me, sister. I promise you I'm quite immune.'

She dug her fingers into her palms. 'I'm not crying.'

'Just as well. I can't bear self-pity.' He gave her a quick glance. 'You're content here, I don't doubt it. You have friends, acquaintances.' She didn't reply. He raised an eyebrow. 'An attachment even?'

Again, she shook her head.

He looked gratified. 'So there's nothing to keep you here.'

'No.'

'And even if there were, we must all make sacrifices. As I have done since Father died.'

'Surely you are free to do as you like, go where you like,' she said. 'Who could stop you? What choice do I have?'

'You know nothing,' he said, coldly. 'There's the small matter of money, for one. None of your concern, but someone has to manage it. I would not have wasted scarce resources on your education, but our mother insisted on it. There it is, it's done. Now you must contribute in whatever paltry way you can. Are we agreed?'

She said nothing, looked out the window.

'Madame? Do you mind what I say?'

'Yes.'

'Good,' he said. 'You'll go to Ballylickey, to be a companion for Mother, and help Jack with the house and lands.'

'As you wish, Emanuel,' she said.

When he'd left she went again to the window. It had grown dark. A carriage passed by, the driver a featureless hulk buried in his cloak. The depthless night sky hung over the rooftops, sprinkled with early stars. As her chest heaved, her breath puffed white, clouding the glass.

❦

Every year, in common with the entire populace, she longed for spring. A thaw that could be felt in bones, fingers and toes. A sweeter, softer gleam to the light; urgent birdsong splitting the air. The tracing of ice on the windowpane becoming gradually less frequent, then melting away entirely. But when Janey came into her room one morning in late February, saying, 'Thank God, Miss, it looks as though winter has done its worse,' and threw back the shutters, letting in a thin sun, Ellen turned her face to the wall and drew the blankets over her head.

'Are you ill, Miss?' Janey's voice was confused, concerned.

'No.'

'What, then?'

'I'll get up in a moment.'

After a pause Janey said, 'Very well, Miss.' Ellen heard the door close. She threw back the blankets, stared at the ceiling.

It couldn't be put off for much longer.

Over breakfast she said to Louise, 'Now that the roads improve, I should give thought to returning to Cork.'

Louise laid down her knife. 'Dr Stokes has said as much. He wanted to wait to tell you closer to the time, so as not to concern you unduly.'

'Has a date been decided?'

'Three weeks, I believe.'

'I see.' She pushed away her plate.

'We would be glad to have you stay longer. You've been a comfort to me, certainly. And the children have grown fond of you. Even Harriet, in the end; she who is slow to warm to anyone outside the family.' Louise dabbed at the corner of her eye, sighed. 'Perhaps, though, you'll be glad to escape the burly of children?'

'On the contrary. I will miss their play and innocence.' Ellen forced her voice to cheerfulness. 'But I look forward to returning home.'

'You're sure?' Louise sounded dubious.

'I'll be glad to see my mother once again.'

'Well, there is that. Still, it's a change from what you've grown used to.'

'I trust I'm adaptable.'

'Women usually have little choice to be otherwise. Have you told Mr Taylor you're leaving Dublin?'

Caught unawares, Ellen blushed. 'Not yet.'

Louise's ice-bright eyes didn't blink. 'He'll be sorry, I'm sure. You seem so … compatible.' Ellen said nothing. 'May I offer some advice?'

Ellen looked at her in surprise. Louise didn't usually ask before offering her opinion. 'Please.'

'Ask him to visit. He is kind-hearted. But men lack imagination, particularly young, popular men at university. So, ask him to come. Besides, he has relations close by, does he not?'

Again, the blood rose to Ellen's cheeks, stinging the skin. 'Yes, ma'am.'

For a moment, Tom didn't respond, just stared at her. The room was full of Dr Stokes's friends and students, talking in informal groups: a soirée of the sort that occasionally turned the drawing room at 16 Harcourt Street into a meeting place of some of Dublin's most important citizens. There stood Mr Petrie, the miniaturist; across the room political leader Mr

Grattan was attended on by Louise. Ellen knew by her vigorously nodding head that she pretended fascination. In fact,
she found these evenings a trial, relaxing only when the front
door closed on the last guest.

She felt his hand on her elbow. He steered her into a corner by the window. 'Why must you go?' he asked.

Now Ellen wished she'd waited for a quieter moment,
though there was no guarantee when that might have been. 'I
told you, my mother needs me.'

'Is there no one else? Your married brother …?'

'No.'

His hand had remained on her elbow; now he dropped
it. He looked around the room, as though hoping someone
would volunteer another response. 'When will you go?'

'Two weeks' time.'

'Two weeks!'

'Dr Stokes will see me as far as Kilkenny.'

'I could accompany you, make sure you arrive safely.'

Her heart jumped. Then she shook her head. Dr Stokes
would never allow it. 'It's kind of you to offer. But my brother
has made arrangements.'

'So there's naught to be done.'

'No,' she said.

'What are your thoughts on the matter?'

'What a sister's, a daughter's should be. Happy I can
oblige my brother and help my mother.'

'I didn't credit you with humbug.'

'What do you mean – humbug?'

'You can't be happy to leave this house. All this.' He indicated the noisy room. 'Dr Stokes has such hopes for you. Now
you are to disappear into the country.'

'We're not all free to decide our future.'

'Granted. But why not at least try and fight for it? Did
you convey your disappointment to your brother?'

'You can't know my feelings.'

'I can guess.'

'Is that not presumptuous?'

'Perhaps. But knowing how I myself would react in your situation, I say you are unhappy. If you explained to your brother ...'

'Then his decision would be the same.'

To her mortification, her voice wobbled. Tom peered at her. The room was stifling. He hesitated, then seemed to deflate. 'Forgive me, Ellen. You're correct, of course. My blood becomes heated and I speak without thinking. You must know this by now. And the truth is ...' He shook his head. 'You are the last person I'd wish to offend.'

'I'm not so sensitive as you imagine me.'

'That's not what I mean. You must know I crave your good opinion.' His eyes shone in the candlelight, with such sincerity, such warmth, she had to look down, away from his gaze.

'You have that, always,' she said.

A call, from across the room: 'Young Taylor!' Dr Stokes, seated, surrounded by a circle of gentlemen, was beckoning Tom to join them. Tom raised a finger – one moment – then turned back to Ellen:

'I'll call on you before you leave, and we can talk further.'

'That would cheer my last days.'

'Will you give me your hand, to prove we are still friends?'

'Gladly.' She held it out. He clasped it, held it, all the while with a grave expression on his face, the most serious she had ever seen him. After a moment, just as she began to wonder if they were drawing attention from others, he let it go, turned and walked across the room. She watched as Dr Stokes introduced him to the group. He bowed, smiled, and was absorbed as one of their own.

☙

'Is something wrong, Miss Hutchins? You start every time the door opens,' Louise said.

'I'm excitable, perhaps, at the prospect of the journey.'

Louise looked at her a moment longer but let it go. 'It happens I have a pleasant diversion for you. We're invited to the Langley's. They have a well-known preacher from England staying, who intends giving an enlightening talk after dinner.' When Ellen didn't respond, her eyebrows rose. 'Does that not appeal?'

Ellen hesitated. 'I have a slight headache, ma'am. Perhaps I could stay home.'

'Alone? To do what?'

'I hoped to finish my book.'

'You mean you prefer to stay here and mope. I consider that ill advised. You would do better to accompany me and take your mind off the next few days.'

'Yes, ma'am.'

'What you'll do in the future without my guidance, I don't know. A young woman needs a mature, experienced friend to coax her along.'

'I suppose I must trust to my own resources from here on. Inadequate though they might be.' This came out sharper than she'd intended; fortunately Louise seemed not to notice for her face softened.

'Well,' she said, 'you're not completely without sense, I admit. Come this evening. You won't have many more opportunities to experience Dublin society.'

At the appointed hour, Ellen rose and followed Louise and Harriet downstairs. They cloaked themselves, tied on their bonnets and waited for the carriage. The journey to Merrion Square was a short one; barely fifteen minutes after setting off they drew up outside an immaculately trimmed town house. Ellen hung back, allowing Louise and Harriet to go ahead and absorb much of the greetings, the approving comments regarding the fairness of the weather, the good health of the hostess. During it all – the drawing room chatter, elaborate and insubstantial as lace; the sipping of tea; the visiting preacher's impenetrable, dreary sermon – Ellen kept her expression neutral and fought down a desire to run back down the stairs and through the front door.

Louise was right, after all. He hadn't called the last two weeks; why presume he would this evening?

At last, it ended. They escaped, returned home along the dark streets, passing tall, elegant houses where candles now winked in the windows.

Louise swept through the hall of Number 16 almost without slowing, handing over her bonnet, patting her hair in place. 'Any messages while we were out, Plunkett?'

'Mr Taylor called, ma'am.' Did she imagine it, or did he announce this with a degree more gravitas than usual? Ellen had given up trying to read meaning in Plunkett's rigid shoulders, a stray twitch of his mouth. As always, he was now staring ahead, beyond expression. She turned away, busied herself with unfastening her gloves. She could not seem to undo the top button of the left, it kept slipping out of her fingers.

'Oh?' Louise slowed and spun around. 'No doubt to see you, Miss Hutchins. Did you expect him?'

'Not as such,' she said, still fiddling with the glove.

'And you leave the day after tomorrow. A shame. Did he leave a message, Plunkett?'

'No, ma'am.'

'Well, it can't be helped. How could we know he would call? He is too casual, that young man.'

The damnable button finally came loose; she peeled away the glove, only to drop it, from tiredness and frustration. It fluttered to the floor, she bent to retrieve it. Plunkett swooped

down at the same moment, she looked up, for an uncomfortable few seconds their eyes met. Did she see there a brief
glimmer of sympathy?

∽

*Tom—I was sorry to miss you. I had hoped to see you
before I left. No doubt you have been busy the last weeks.
Providing it does not intrude on your studies, I should
be interested to hear how you progress, at botany and all
other matters.*

She paused, holding the grey goose feather over the page,
then scribbled on:

You are ever a stimulating friend.
E. Hutchins

She sealed the note, put it in her pocket to give to Plunkett
after breakfast. Yet when the meal had finished and the servant bowed and asked, 'Is there anything else, ma'am?' she
stayed silent. Louise dismissed him with a wave.

Once in her room Ellen folded the note between the pages
of her bible. She feared her carefully chosen words showed disappointment. She would not let him think he owed her anything.

∽

Dr Stokes found himself talking through dinner in an exag-
geratedly hearty manner, mostly into silence. Miss Hutchins
hardly spoke. Louise had been no help. Was she so attached
to the girl? He hadn't realised. Sentimentality: contagious as
any fever, the compelling desire to show solidarity manifested

through a physical outpouring. He observed it frequently in his female patients.

'Are you prepared for the morning?' he asked.

Miss Hutchins rallied, sat up straighter. 'Yes, sir, I've packed. Mrs Stokes kindly helped.'

'We went shopping for a few necessities earlier,' Louise said. 'Who knows when Miss Hutchins will next be near a good haberdashery?'

'They have shops aplenty in the city of Cork, I believe.'

'She does not go to a city, but to the middle of nowhere.' Dr Stokes raised his eyebrows. He and Louise had discussed this earlier.

'She will no doubt be able to send for anything she needs,' he said. 'And Bantry has a good market day, or so I hear.'

Louise threw down her napkin. 'A market will not compensate. A young woman needs society. For example, where will you attend church, Miss Hutchins?'

'There is a fair-sized church in Bantry, ma'am.'

Louise's eyes narrowed. 'How far from Ballylickey?'

'Three miles or so, I think.'

She blazed in triumph. 'Three miles! I knew it! And how is she to get there in the middle of winter, I ask you? Along a country track?'

'Presumably the same way Mrs Hutchins gets there currently. Really, wife, you make it sound like some wilderness. You forget that Miss Hutchins spent her earliest years there, and her family for generations before that. The life is in her blood.'

'She was but five years old when she was last there,' Louise said. She had clenched her fist in its lace glove. 'Now, just when she begins to come out of herself a little, to be at ease amongst us, she is sent away.'

'Of course, we will miss her,' Dr Stokes said. 'But as Mr Hutchins explained, it cannot be helped. Miss Hutchins must do right by her mother and brother.'

'I know it, sir.' Ellen turned to Louise. 'I am quite resigned to it, ma'am. Please don't grieve on my behalf.' She studied her plate. 'Though I appreciate your display of feeling.'

This is unsettling the girl. He should draw a line under the discussion.

'Change is never easy,' he said. 'We resist, and only cause ourselves unnecessary discomfort, particularly when it is thrust upon us, rather than actively sought. But we grow by accepting life's twists and turns, not defying them. I have every confidence that Miss Hutchins will demonstrate the resilience I have observed in her, and in doing so justify our pride in some way facilitating the development of an exemplary young woman.'

He'd intended this speech to be encouraging; now it sounded sham even to his own ears. Worse still: to his horror, the girl's eyes grew misty. He cleared his throat. Best to act as though he hadn't noticed. 'Besides, you will always have a home in Dublin, should you ever be in a position to return. And we hope you will at least visit us.'

'Yes,' Louise said. 'You must certainly visit us.'

Miss Hutchins's face brightened, then clouded again. 'The distance is great. They may not be able to spare me.'

'The roads improve all the time,' he said. 'And you are far stronger now than before. Your mother could accompany you. All manner of possibilities present themselves when we remain open to better fortune.'

Later, before they retired to bed, he called her aside. There might not be time in the morning. 'I have something for you.' He indicated the books where he'd laid them out on a side table: a small stack of four volumes that made up the whole work. She lifted one, flipped the cover open to the title page.

'*A Systematic Arrangement of British Plants,* by William Withering. Can you spare this, sir?'

'I have another set. I want you to have this one,' he said. He took a box from his vest pocket and unfolded it. 'This comes with it.'

'Oh, sir.' She flushed – as was her habit – at the sight of the small microscope.

'It's a basic model, but enough to get you started. See, place the plant sample here on the stage and turn this screw to focus the glass.'

She frowned, following his hands.

'The instructions are printed here, on the inside of the lid. It's simplicity itself. There are other useful bits and pieces included – a forceps, knife and needle. And a separate small glass for quick studies.' He closed up the box again, placed it in her hands.

'Thank you, sir. But will I have enough use for it? I have not your great knowledge.'

He gestured towards a chair. 'Sit, please. I have something important to say.' She did as he bid. He sat down opposite. 'Mrs Stokes is correct in one respect. Where you are going is isolated. There will be few visitors.'

'I remember.'

'I wonder if you do. A child has a different perspective. Less of a need for society, for one thing. I notice you tend to pensiveness. That is not a bad thing in itself; it indicates a sensitive disposition. But looking inward excessively will lead to melancholy. Do you understand?'

'I think so, sir.' Her voice was small.

'Another thing. Your physical health. It has improved greatly since you came to us. But you must be ever vigilant, and continue to build your strength.'

'Yes, sir.'

'On both counts, I recommend you take as much exercise as possible. Walk, Miss Hutchins, walk.' He struck his fist against his knee for emphasis. 'Let neither rain nor wind stop you. Harden out, as young plants must be hardened out in early spring, unaccustomed as they are to sharp breezes and night frosts. I prescribe this as your physician – I ask you to take it seriously.'

'I will, sir.'

'Very good. Now, as to the study of botany.' He pointed to the book, which she clasped in her lap. 'You need a pastime to divert your intellect. Therefore when you are out on your walks, exercise your mind at the same time. Collect whatever plants catch your eye; examine them under your microscope. Locate them in Withering and study what he says closely, but always make your own observations. Make the landscape your own. Turn your situation to your advantage. Withering says it himself, *... the Study of Botany, in particular, independent of its immediate use, is as healthful as it is innocent. That it beguiles the tediousness of the road, that it furnishes amusement at every footstep of the solitary walk, and, above all, that it leads to pleasing reflections on the bounty, the wisdom, and the power of the great CREATOR.'*

She ran her hand across the golden brown binding. 'I may not have a talent for it.'

'You're thoughtful, intellectually curious, able-bodied. You have a keen eye. A lady in your position should find some leisure hours to fill. Not much more is needed. A vasculum, perhaps, and a drying box.'

'Is it ... suitable?'

He frowned. 'Suitable?'

'What I mean is ... would my brother have any objection?'

He tutted this away. 'There can be no objection. It's the most natural of activities, utterly appropriate.' Seeing her creased brow, he added, 'I shall write to Emanuel if needs be.'

She said nothing. Across the room, Louise sewed diligently, pointedly not-looking, her lips pursed.

'Well,' he said. 'What say you?'

'I thank you, sir, for your gift and your advice. I will make use of both.'

He leaned back in his chair. 'There is no knowing if you will take to it. In the end, the subject may tire or bore you. You may find it too demanding, too rigorous.'

She lifted her chin. 'That would not stop me.'

'Very good.' Glancing over at Louise, he stood. 'I have lectured enough for one night. You should get to bed, in order to be fresh for your journey.'

She rose. 'Goodnight, ma'am.'

'Goodnight, Miss Hutchins.'

When the door had closed, Louise said, 'Do you think looking for plants will stop her losing her reason in that backwater?'

'It may do.'

'I hope you're right.'

'She will adjust.'

'She should hope instead to make a match as soon as possible. It's all too easy for an unmarried daughter to end up a slave to her relatives. Did it not nearly happen to me?'

'A match is less likely to happen now.'

'Exactly my point. You should have said more to dissuade the brother.'

A flare of irritation: it wasn't Louise's place to question him. He bit back a harsh reply, in part because he feared she might be right. 'I can't dictate to Emanuel Hutchins how to manage his family affairs. Besides, he is quite resolved. I've told you how rigid-minded he is. The subject of his family seems to make him unspeakably tense. I could not get him to discuss his sister in any kind of open manner.'

'Even though you have nursed and fostered her all this time.'

'Even so.'

'I don't understand your regard for him.'

'I've told you. He showed ... loyalty, when others did not. He always spoke in my defence. When my very liberty was at question.'

'As you did for him.' Louise's eyes flashed. 'He is too much of a sneak to ever have been at risk himself.'

'Louise. You go too far.'

She raised her chin. 'I hold to my opinion.'

'Even when I ask you to temper your feelings.'

Now her mouth worked. 'I hate to see that girl waste her life.'

'You can't know that will be the case. It's not a foregone conclusion.' He held out his hand. 'Now, let us go to bed as friends. And try and part from Miss Hutchins tomorrow with good cheer. She will take her cue from you.'

She put her sewing aside, slightly mollified. 'Yes,' she said. 'There is that. Older and more used to suffering and loss, I must lead by example.'

The morning of her departure, Janey shyly handed her a small package.

'What is it?' Ellen asked.

She blushed, looked down. 'The caul from Master William's head when he was born, that covered his face. I was told to throw it in the fire, but I couldn't. It has special powers.'

Ellen almost dropped the piece of folded linen. 'I can't take this, Janey.'

She looked up, eyes wide. 'You must, Miss. It brings luck. They say you're going to a terrible wild place. I want you to have it.'

She tried another tack. 'You cannot part with so precious a thing. You'll need it for yourself.'

'It will ease my mind, that naught will happen to you on the journey. Keep it safe. Please. Promise, Miss.'

Her face glowed with sincerity and good will. Ellen promised. Now she was bound to the horrible thing.

᭣

Plunkett tripped past her down the steps, directing the loading of the trunk. Louise embraced Ellen goodbye. Her manner was queenly, brave, resigned. She called to Plunkett over Ellen's shoulder to *please take care of Miss Hutchins's things*, while smartly returning the nod of a lady of her acquaintance, gliding by on the pavement. Through the open front door came the sound of children's laughter and a baby's distant wails.

Ellen wished they could be away. Nothing left to say, the rituals of parting complete.

Life at Number 16 already carried on, even before she stepped into the carriage. Oh, in the future they might ask: how many years since Miss Hutchins left us? Finding a button on the floor – she'd lost one from her summer cotton – they might exclaim: did that not belong to Miss Hutchins? Or recall with amusement something she'd said; though to her recollection she'd never said anything very amusing, or interesting.

Was that her fate? To become the slightest of footnotes in their crowded family history?

Passing Trinity College on the way to the coach stop, she couldn't stop herself from glancing across at the groups of young men that loitered around the gates, or ambled together through the great arched door, disappearing into the courtyards beyond. She recognised no familiar stride or set of the shoulders, no eager, resolute face beneath its crop of shining dark hair. And if she had, she could not have called out, or followed him, but must turn away, as she did now, to her own course.

BALLYLICKEY, CO. CORK
MARCH 1805

FIVE

Indigo streaked the western sky.

The house was fading into shadow, consumed by the mass of tall trees behind. Rooks flew, screeching and thrashing, into the high branches. Ellen unfolded her aching limbs. Dan, the Hutchins's manservant and driver, helped her down from the carriage.

'Ellen.' Leonora Hutchins came forward from the open doorway, a wispish figure wrapped in black silk.

'Mother.' How insubstantial she was, as though made of straw. Ellen feared the embrace would crush her. Still she held on. 'I thought we should never get here,' she said. 'The road from Cork seemed interminable.'

'The last miles are the worst,' Leonora said.

'A section of road was completely bogged; your servant had to cut and lay branches across before we could pass.'

'You're fortunate to reach here before dark. It's no place to be stranded overnight.'

A dragging sound across the gravel: Jack Hutchins had come from the carriage and now joined them, leaning on a crutch.

'Let's continue the reunion inside,' he said. 'A chill has settled in my bones and I need the fire.' His thin face was white and drawn with exhaustion. When he had met Ellen off the coach in Cork City, he had seemed at once familiar and strange. She recognised something like her own features, but hardened, a twist to the mouth, deep grooves in his forehead. He looked undersized for his age – twenty-seven – but wrenched her fingers in an assured handshake. He spoke little on the day-long journey to Ballylickey, though his eyes darted to her frequently, as if in fascination.

They went in. Jack slammed and bolted the door behind them, limped away down the hall. Ellen stared about, her eyes adjusting to the gloom. The candles had not yet been lit. 'It's smaller than I remember,' she said.

'Naturally enough, after all this time,' Leonora said. 'You were but a little thing when last you were here. Ah, here's Kate.'

'Miss.' The servant took Ellen's coat and bonnet. She looked around thirty, small, with a broad mouth and freckled cheeks.

'Kate runs the house. We couldn't manage without her.'

'You'd surely get by, ma'am,' Kate said. Her eyes met Ellen's, briefly: a clear hazel-brown. 'I'll put away your things, Miss.'

'We have few servants these days,' Leonora said. 'None that were here when you were a child. The household is greatly diminished. There's Kate and Joanna for the house, Dan Murphy for outside. A lad for the stable, a few farmhands.' She gestured towards an open door to the right. 'There's a fire in the parlour, you can both take your dinner there.'

'I'd like to wash my face and hands first, if I may.'

'I forget your long journey. Kate!' She emerged again from the shadows. 'Take Miss Ellen upstairs to her room.'

'Yes, ma'am.'

On the way upstairs Ellen ran her hand along the carved wood of the banister. Smooth, worn warmth: this she remembered. The pictures – stiff men and women clad in black; muddy, brown landscapes – also stirred her memory. She had known these flat faces, gazed up at them in awe.

At the top of the stairs, she pointed down the long, dim landing. 'I suppose I can't sleep in the nursery any longer,' she said.

'No, Miss,' Kate said, pushing open a door. 'That wouldn't do at all.'

A large, bare room, furnished with a bed, a chair, a chest of drawers. The fireplace was handsome, of pale pink marble. Two large windows framed a view to the sea, the bleary sky.

'Your mother says you may choose any furnishings you like from the other rooms,' Kate said. She folded her hands across her stomach. 'No one has slept here for many years. Shall I draw the shutters? It will be dark soon.'

'Leave them open for now, please.'

'There's clean water in the jug.' The servant crossed to the door. 'If I can do anything for you, Miss, please call.'

'Thank you,' Ellen said.

Kate had one hand on the doorknob. She said quickly, 'We're all glad you've come, Miss.' She went out and shut the door before Ellen could say anything.

'The parlour hasn't altered very much.' Leonora stood at Ellen's elbow. A circle of chairs and a long sofa heaped with cushions and blankets crowded the space before the fire. The walls were clad in cream paper, tattered and peeling in places, and hung with sepia seascapes. In a corner, a stuffed fox crouched in a glass cabinet.

'That creature I remember,' Ellen said. Its wild, glassy stare and toothy smile had stalked many of her night terrors. Now its fur was patched and faded, its frozen grimace pathetic rather than frightening. 'And Jack, you used to climb onto that dresser and jump onto the sofa.'

'Did you, Jack?' Leonora said. 'I don't recall.'

'I was a daredevil,' he said, 'then.' In the short time she'd been gone, he'd already eaten, the remains of his supper nothing but a greasy lick on an empty plate. He slumped in an easy chair nearest the fire, cradling a wine glass.

Leonora gave Ellen a wild look. 'Well, my dear, sit, eat something. You're surely starved.'

Cold mutton and potatoes, bread. She'd spent the last hours of the journey dreaming of food. Now her appetite had faded away.

'Mother,' Jack said. 'Sit. Don't gape at Ellen like she's a rare animal.'

Leonora sat on the edge of the sofa, tucked her hands into her lap. Her eyes were bright in the firelight. Ellen picked at the mutton, managed a fork of potato.

'I can have Kate fetch something else,' Leonora said. 'Cheese, or jelly.'

'Don't trouble yourself, please.'

'You must let us know what you like. Though we have few luxuries, Kate is a competent cook.'

'I prefer simple food. And this is good mutton.' Ellen pushed her plate away. 'But it seems I'm too tired to eat.'

'Tomorrow you will be brighter,' Leonora said, 'and can view the house and lands.'

'I'll give you a tour,' Jack said.

'But how ...' Ellen stopped.

'How do I get around? Ah, you'll see.'

'Jack manages very well,' Leonora said.

'It takes more than a working pair of legs to run this place. There's naught wrong with my mind.'

'That's obvious,' Ellen said.

He smiled. 'You might be useful to me, sister. In all manner of ways. I need someone I can trust. You've had some schooling, at least.'

'Some.'

'My own was interrupted, as you know. By calamitous misfortune.'

'I know it.'

'It doesn't help to dwell on matters,' Leonora said quickly. 'You were spared, thank God.'

Jack leaned back in his chair. 'Did you pray for me in that school after I broke my back?'

'Day and night. As did the rest of the pupils, and the teachers.' *That Miss Hutchins's brother might recover from his accident on the ice and regain the use of his limbs, we beseech thee, God. Amen.*

He held up his glass, in a mock toast. 'Thank you for that, at any rate.'

He appeared on edge. Was this the reason for Leonora's apparent nervousness? She changed the subject. 'Shall I see Arthur soon?'

'He's extremely busy with the estate,' Leonora said.

'His demesne, by the bay,' Jack said. 'He's planting a forest and is utterly consumed by it.'

'What is his wife like?'

'Matilda?' Jack shrugged.

'A sensible person,' Leonora said. 'Practical. She suits him well.'

Silence. The weight of their long years of separation seemed to hang in the air.

'Seeing you here is like a dream, Ellen,' Leonora said, suddenly. 'Is it not, Jack? She is surely a vision.'

'She's corporeal enough. Touch her, she will prove flesh and blood.' Taking up a folded newspaper, he threw it at Ellen. It smacked harmlessly against her skirts and flapped onto the floor, revealing the front page of *The Cork Mercantile Chronicle*.

'Have you lost your senses?' she cried.

'See, Mother? She is real as you or I.'

Leonora raised her hands in a gesture of weak dismay. 'Did he hurt you, Ellen? He means nothing by it.'

Jack's mouth hung open in a slack grin: a boy's satisfaction at his own cleverness – *I got you!* His eyes glittered. 'We don't indulge ladies here,' he said.

'I've been far from indulged, I promise you,' Ellen said. 'And no, that did not hurt, the throw was too feeble.'

Jack sat up straighter, delight in his eyes. 'I can try harder,' he said.

'Enough, Jack!' Leonora said. 'Ellen is used to refined manners. She'll take the first coach back to Dublin if you continue to misbehave.' Her mild tone took the sting out of the admonishment, and Jack relaxed again in his chair.

'I've been living amongst little children,' Ellen said, 'and have grown used to their attention-seeking ways. Ignoring bad behaviour usually ends it.'

Jack laughed, a dry bark. 'I perceive Ellen has a sense of humour,' he said. 'That will make life more interesting. What do you do for amusement, sister? Do you play the pianoforte – not that we have one – sing, play cards?'

'I like to read.'

'How very dull. That newspaper by your feet is the extent of my own reading habits.'

She glanced quickly around the room. There was a bookshelf, with a collection of volumes that she might get through in a month. Apart from that, a couple of old almanacs on the dresser, a stack of newspapers in a corner. 'I'll need more books. And I intend on doing a lot of walking.'

'Walking?' Jack shook his head. 'In these parts?'

'Certainly.'

'Where?'

She waved her hand. Everywhere. 'The woods, the beaches, the mountain tracks …'

'You'll find it hard going,' he said. 'For one used to strolling down Grafton Street. The paths around here – where there are paths – are quite different. Hazardous, rough. And you'll be alone, which is another matter. God knows I can't accompany you.'

'It's not safe, Ellen,' Leonora said.

'I'm used to rambling,' she said. Their faces were skeptical. 'I intend starting as soon as possible.'

Silence. The fire crackled in the grate. 'We'll speak of it again,' Jack said. 'Tomorrow, I'll show you the lands.'

⋘

A low seated, two-wheel, gig-like contraption stood hitched to a pony in the yard outside the back door.

'My replacement legs,' Jack said. He patted the pony's flank. 'Meet Jacob.' Murmuring to the animal, 'Steady, steady,' using his crutch as leverage, he hoisted himself twisting and sliding into the seat and took the reins. 'Come.'

She hesitated. The gig looked as though the first rut in the road might flip it over.

'I hope you aren't of nervous disposition,' Jack said. 'That won't serve you around here.'

'I'd rather not break my neck,' she said.

He raised his eyebrows, smiled. 'Touché, sister.'

'I didn't mean ...'

'I know.'

Already she noticed a pattern to his moods. Today the fissures in his face had smoothed, as if thawed. He tossed hair from his eyes.

'We'll take it slowly,' he said. 'You'll see it's quite comfortable.' He held out his hand. Ellen lifted her skirts and sprang onto the step, allowing him to pull her up beside him.

For a while he stayed true to his word, keeping Jacob to a careful trot, following well-worn grooves along a track into the fields. He pointed as he went – 'Potatoes from late spring. Wheat and barley, come summer. Dry hay for winter fodder. And those fields I'll leave fallow, for next winter and spring planting.'

In the next field over the hedgerows a cluster of men stooped, hacking at the soil with spades and digging out stones with bare hands. All wore some kind of footwear, though their ragged trousers and thin jackets whipped in the wind. A cluster of crude dwellings, no more than cabins of packed earth and straw, bordered the plot. Grey smoke dribbled into the cold sky. Children ran through the dung heaps.

'Most of the land is leased,' Jack said. 'I manage it myself, don't use middlemen.'

'You never have trouble?'

He shrugged. 'We're known for being good landlords. That is, I try to be fair. I have sympathy for the tenants; it's not an easy existence. There's a bailiff if I need him. But it rarely comes to that.' He waved his hand. 'Good day,' he called. The men straightened their backs; those that wore caps removed them. 'They speak Irish amongst themselves,' he said aside to Ellen. 'But there's always one at least in a group who'll have some English. Usually not the women or children, though.'

'What do they live on?'

'Potatoes. And little else, poor devils.'

He clicked Jacob on. Having reached the boundary, they tracked the edge for a time. Ellen glanced up. The land lay as a valley amongst shores of rock, rising into bare hills tinged with cream and yellow.

'Are those ours?' she asked, pointing to a scattering of straggly sheep dotted across the upland.

'Yes,' Jack said. 'Hardy, self-sufficient creatures. We also have a few cows, for our own use, mostly. Pigs. Fowl – hens, ducks, geese – you may help with those, if you choose.'

'I want to be useful,' she said. 'Though I have no experience.'

'There's only one way to acquire any, and that's by doing,' he said. 'Besides, there isn't much to learn. Fowl aren't complicated. They eat, defecate and lay eggs, all day, every day if you're lucky. Occasionally we have one for the pot, if it's grown too old. Then watch it run around the yard without its head, spouting blood.'

Ellen said nothing. No doubt she'd get used to his careless way of talking.

'Well,' he said. 'That's the extent of our kingdom. What think you of Ballylickey?'

'I'm no judge of such things.'

'You have two eyes, and a tongue. And opinions, I'm sure.'

'Then I think it very fine,' she said. 'We seem quite self-sufficient.'

'Fortunately, as we're so remote. We live adequately, though there's never money to spare; we cut our cloth to fit our measure.' He peered sideways at her. 'No doubt you thought badly of coming down here after living the high life in Dublin.'

'The Stokeses are modest people and lived accordingly.'

'But cultured, more refined than we are?'

'I won't deny that.'

He snorted a laugh and drove on, waving flies away from his face. After a minute he said, 'Mother says you speak French.'

'I won't have much need of it down here.'

'Even here, there's need of French spoken, in certain circumstances. You know Father was fluent in the language?'

She looked at him, surprised. 'No. Or if I knew, I'd forgotten.'

'The reason perhaps, you were sent to that school.'

'He was a cultured man, it seems.'

'Aye, he was. Though he used his French mostly for dealing with the smugglers.'

'Those that came before him, you mean, as magistrate?'

'No, sister. I mean the smugglers he bought brandy from, and sold on.'

'You're amusing yourself, Jack, at my expense.' She tried to keep the primness from her voice, but couldn't help adding, 'And that of our father's memory.'

Now he laughed freely, fully. 'Come down from your high horse, sister. It's true enough.' She said nothing. 'Oh, he wasn't involved in a serious way. A small player, I'm told. An opportunity presented itself from time to time and he took advantage of it. There's a bottle or two around still. Superior fire water.'

'I don't believe you.'

He shrugged. 'Do or don't, I care not.'

After a minute she said, 'Is this commonly known?'

'If it is, it shocks no one. The line between law and transgression is less defined in this part of the world. Even a gentleman may take a certain license from time to time.' He leaned into her shoulder, as if to nudge her from her disapproval. 'It was many years ago, sister. The Hutchinses no longer lurk about the beaches at midnight, to cart away crates of the best French brandy, or any other contraband.'

'No?'

'Not I, at any rate, for obvious reasons. Arthur is too conscious of his role as local squire. Emanuel might have the nerve, had he the inclination. But he is far from here, most of the time. So we lead a quiet, law-abiding existence, as dull as you might wish.'

'I don't ask for a dull existence. Simply a peaceable one.'

'It's the monotony you must watch for, sister. The long, drawn-out dark of January, when every tick of the clock cracks against your ear like gunshot. When you long to howl just to break the silence. These mountains …' His voice fell away. 'We must take our amusements where we can, or lose our minds.' He threw out the reins and clucked at the pony to go on. It broke into a canter, then at Jack's encouragement, a clumsy gallop. Ellen gasped. The gig bounced across the rutted track, rising into the air as it hit against rocks and hillocks. She gripped the sides, trying not to shriek. From the headlong, mad, loose thrill of it, and the terror.

She took a large rug in shades of rose from the parlour – hardly missed from the clutter – and a selection of framed prints from around the house; flower arrangements, a view of Cork harbour that reminded her somewhat of Dublin, miniatures for the dressing table. From the seldom-used back drawing room she took the curtains; her own were frayed, patches of light leaking through the thin fabric.

Kate helped hang them. Standing back, she said, 'Yellow is cheerful, Miss.'

'The light in the drawing room is so poor the silk has been well preserved.'

Kate stopped in front of a framed silhouette of a woman's head, cut skilfully from black paper and mounted on ivory card.

'Pretty, is it not?' Ellen said.

'That it is, Miss.' Still, she stood there.

Ellen cleared her throat. 'Was I right to take it?'

'As you wish, Miss. It's just … do you know who it's of?'

'Some ancestor, I presumed.'

Kate nodded. 'In a way.' She paused. 'Your sister made it.'

A prickling ran along the nape of Ellen's neck, where hair met skin. 'Katherine?'

'Yes, Miss. A portrait of herself. They say around here that she was very handsome.'

'I wish I could remember her face.'

Silence. Then Kate said, 'Your mother says this was Miss Hutchins's room once upon a time.'

Ellen started. 'I'd forgotten.' After a heartbeat, 'So she must have died here.'

'I can't say, Miss,' Kate said. 'I wasn't here then. Most likely, though.' Curiosity overtook her usual guarded manner. 'Do you remember anything of that time, Miss?'

'Nothing. Though at four, I might have retained some impressions. All I recall is a fog of horror that seemed to hang over the house in the months before I went away to school. It's how I thought of Ballylickey, for years after.' She wound her arms about herself, suppressed a shiver. 'She fell from a horse a month before, that much I know. Whether that caused her death could never be proven.'

Why was she telling the servant this? She never spoke of it, even to Caroline. Something about Kate's presence: solid, as though nothing could shake her.

∽

A sullen sky, opaque and tinged with lavender. The bay, contained as though in a basin, merged with the horizon. She had come to the shore hoping to follow it as far as Arthur's estate farther down the coast, setting out from the family boat house that slumped on the edge of the bay. After twenty minutes of slow progress she admitted defeat, thwarted by a clump of impenetrable, shrubby trees. All she had for her pains was a scratch on her cheek. She turned inland towards the Owvane River and resolved to follow its banks. Here the going became easier, and she walked entire stretches unimpeded. The breeze blew cool and fresh; the fields were lushly green and devoid of a human soul. She picked up speed, getting warm, her limbs growing used to the swing and stride of her walk. Passing close to a settlement – visible as a cluster of smoke plumes – she paused. She preferred not to meet anyone.

She should get a stick, a sturdy stick. And a dog.

Something caught in her slipper; she stumbled. Stooping, she undid the lace and tipped out a stone, hoping no one would come upon her while so exposed. She added boots to the list: leather boots with stout soles to shield her feet from wet and injury so that she could walk where and when she wanted. These slippers were more suited to a drawing room, or a stroll on the neat gravel path of a gentleman's garden.

When Kate came into the parlour later enquiring about din-ner, Leonora said, 'You must speak with Miss Ellen about such matters from now on.'

Kate turned to Ellen. 'There's a hare, Miss, caught this morning.'

'I see,' Ellen said.

'I could boil a stew out of it, with turnip and barley.'

'Whatever you think, Kate.'

'I'm making butter later, there's cream. And we need supplies from town. Flour and such.'

Ellen nodded. 'Very well.'

Kate waited as if to say something else, nodded, bobbed – 'Thank you, Miss,' – and left. After a moment, Ellen got up and followed her down the back hall.

The whitewashed kitchen was long and low, and reeked of grease and yeast. The scullery maid Joanna had been scrubbing the long table and her mouth hung open with effort. Seeing Ellen, she stopped. Wet trickled down her bare arms. She blew a stray hair from where it had fallen over her cheek.

'Kate.'

She turned from the table, straightened. 'Miss?'

'It seems my mother wants me to help with …' She looked around at the walls of open shelves, stacked with crockery, the enormous open fire with its roasting spit and suspended kettle and pots. 'With this. The house, I mean.'

Kate said nothing, waiting.

'I know nothing of managing a household. I hope you'll guide me. I'll have to rely on your judgement until I'm more familiar with how you do things.' The girl Joanna looked between them, her flat face shining.

'That's only to be expected,' Kate said, 'With you being away for so many years.'

'Not much of what I learned at school is useful to me now.'

'You've learned to be a lady,' Kate said. 'That much is obvious. And this house needs a lady for its mistress, as every fine house does. Your mother will want to rest now that you've returned.' Her gaze was steady. 'It's not my place to say it, Miss, but she isn't strong.'

'I see that,' Ellen said. 'She can depend on me, from here on.'

Kate dipped her head. A confidence had been exchanged, some kind of pact reached. 'Between us,' she said, 'We'll manage fine.'

'Well, then,' Ellen said. 'Let us go through the pantry and make a list of what's needed.'

SIX

She found Jack in the shed. 'Blast it,' he said as she went in, 'What do you want now?'

He sat on a low stool, holding a short, fat iron bar in one hand. His face was slick with sweat; his shirt gaped open. To her amazement and horror she saw that he wore what appeared to be women's stays around his torso. He quickly pulled his shirt across, scowled: 'Well?'

She stood there a moment, feeling uncomfortable; she'd obviously interrupted some private, intimate act. 'Mother asks if you're coming to Arthur's with us.'

He laid down the bar, took a cloth from where it had been thrown on the ground, wiped his face and hands. 'You're welcome to the journey. My back would seize up for a month after.'

'Very well. I'm sorry to have interrupted.'

'I'm exercising my arms. The only strength I have left is there. I must preserve it if I'm not to be totally useless.' He threw down the cloth, began curling his arm up against his chest, down again, an expression of grim endurance on his face.

He must wear the stays to support his back, she thought. She'd remember it, let pity temper her feeling the next time he irked her.

❧

The road to Ardnagashel wound close to the coastline. To the right, barren mountainside, sparse trees. To the left, patches of green fields amidst woodland.

'Would it not be easier to go by boat?' Ellen asked. Her bones rattled, her brain jostled in her skull. Reaching overhead she found the strap, and clung on.

'I get sea-sick,' Leonora said, raising her voice over the din of wheels bouncing off rock.

'Surely it could be no worse than this,' Ellen said.

'At least I know I'm on dry land. Alas, the condition of the road prevents our visiting Arthur as often as we'd like.'

With a lurch the carriage slowed and turned off the main road. They passed beneath a monumental stone arch: the estate entrance. 'Imposing,' Ellen said. 'Almost feudal.'

'Arthur intends establishing a presence here,' Leonora said.

This road had been laid with gravel; the shaking and rattling eased. 'It's even more remote than Ballylickey,' Ellen said. No signs of habitation: no smoke, no workmen along the road.

'He owns all this,' Leonora said, following her gaze. A vista of bleakness, clearings interspersed with patches of fenced off saplings. After half a mile Dan clicked and murmured to the horses and the carriage slowed to a trundle. There was a sudden, scattered noise on the carriage roof: 'What on earth is that?' Leonora clutched her cloak about her and shrank back into the seat. Ellen looked out the window, in time to see the back view of a small boy disappearing over a wall.

'Our welcome party,' she said.

'What can you mean?'

'Nothing to fear, Mother. A gust, scattering leaves. Or stones, thrown up from the road.'

Now the bay appeared to their left, still and smooth as glass. The carriage rolled to a stop. As they got down, Arthur came from around the side of the house and called a greeting. Servants rushed to the horses. Stepping down, Ellen took in the setting. The strip of road ended here, separating the house a scarce five yards from the beach and sea. The house had a plain facade, unadorned, with angular gables and tall narrow windows, white-washed to a stark simplicity. Still, the overall effect was impressive: the hulking building, assertive in its proximity to the sea, as though it feared neither tide nor storms. A woman stood in the front doorway. She bustled forward, hands held out.

'Matilda.' Leonora put her face up to be kissed.

'Mother. And is this my sister Ellen?' Her smooth face broadened into a smile, no less pleasing for displaying overlapping front teeth. Ellen had an immediate impression of freshness, energy concealed. 'I can scarcely believe we have not met before this.'

'Nor I,' Ellen said.

'We must be great friends from here on.'

'I agree.'

Arthur had been standing by; now he moved forward and Ellen had a chance to look at him properly. From Jack and her mother's talk of him she'd expected a man of vitality and vigour; she blinked, rapidly readjusting the picture she'd assembled in her mind to accommodate this thin, stooped figure, a gaunt face dominated by a large, if well-shaped, nose, looming over a rather pinched mouth.

He stepped closer and kissed her on the cheek, his lips fleeting and dry against her skin. 'Sister. We find you well?'

'Very well.'

'How do you like Ballylickey again?'

'I'm glad to be home.'

'Has it changed?'

'It seems at once smaller, and larger, than it did when I was a child.'

He studied her. 'Smaller, I understand. But larger?'

'The sky is huge, almost overwhelming. And the sea, the land ... I'm unaccustomed to such open spaces.'

'When we first moved here,' Matilda said, 'I struggled with the silence, the endless vistas. A fox barked in the night and I would start awake in terror.' She shivered, shrugged, smiled. 'You'll become used to it, as I did.'

'A period of readjustment is to be expected,' Arthur said. 'It's been many years. I must say you've grown to be handsome.'

'Thank you, Arthur,' she said, taken aback, pleased.

'She resembles you, Arthur,' Matilda said. 'Do you not think so, Mother?' Ellen thought Arthur looked to be under strain, if not ailing; whatever about sharing his features, she hoped she presented a healthier appearance.

Leonora pursed her lips. 'I can never see likenesses between my children.'

'Oh, I think it marked,' Matilda said. Just then two young children appeared: a boy, whippet thin with narrow features, and a girl, all dimples and plump white limbs. 'Now these are as unalike as siblings can be. Freddie, Margaret, say hello to your Aunt Ellen.'

Margaret smiled without affectation and came forward to be admired. The boy stood where he was.

'How do you do, Margaret?' Ellen gave the girl her hand.

'How do you do.' Her voice clear and high, precise as a sharpened pencil.

'Freddie, dearest, won't you say hello?' Matilda said.

Leonora tutted. 'Come child, do as you're bid.'

'Poor boy. He's shy,' Matilda said.

Ellen thought Freddie about as shy as a young turkey cock. 'A strange thing happened as we arrived,' she said.

'Oh?' Matilda said.

'A curious noise, on the roof of our carriage, as though someone had thrown stones at us.'

'Goodness. Surely not.' Matilda looked at Arthur. His face remained impassive.

'It was of no consequence. Of course, someone throwing stones could startle the horses and cause an accident. But I'm sure it was leaves, or pine cones, blown down by the breeze.'

'Most likely,' Arthur said. 'We don't tolerate stone throwers at Ardnagashel. In fact, if we catch them in the act, we give them a good whipping.' Two rosy patches had appeared on Freddie's cheeks.

'No need for that, surely,' Ellen said. 'A reprimand, perhaps.'

'Well,' Matilda said, looking between them, 'fortunately no one was harmed. Won't you come inside and see the house?'

Arthur pointed out the eight-foot-high walls that surrounded the boundaries and divided areas of planting.

'Initially I had problems with invading cattle – my own, and my neighbour's. Expensive young plants, carefully, laboriously planted, lovingly watered … destroyed overnight, trampled, decimated.' He beat the ground with his stick, as if reliving the fury, the impotence. 'And so I had to build these walls.'

'What size is the estate?' she asked.

'Three hundred acres,' he said. 'I intend on planting more trees, in time – beech, firs, Scots pine – and laying out pasture land. Some day this corner of the peninsula might be famous for its gardens.' His voice, up to now precise, almost fussy, had taken on a note of pride, repressed excitement. Touring the house she'd noticed the titles of some of the books, left to hand on a round table – *A New System of Agriculture*; *A Practical Treatise on Planting and the Management of Woods and Coppices*; *Treatise on Agriculture*; *A Compleat Body of Husbandry*.

'It seems a momentous task.'

'I don't view it as such. I feel strongly motivated to accomplish my ambition. A desire to populate what is barren and desolate. To conquer nature, tame it to beauty. Some impulse to transform my surroundings.' They entered an area of more established trees. Between the dense trunks the air had fermented, warm and moist. Ferns and shrubbery filled the spaces between. The ground crackled underfoot, thick with a carpet of dead leaves and seedpods.

'It will be magnificent,' she said.

He nodded: 'Nothing gives more satisfaction than a garden, large or small, be it ornamental or vegetable.'

'I know nothing about plants. Of the type you mean, that is – trees, shrubs. I know a little of botany, from living with Dr Stokes.'

'Some of it rubbed off on you?' he said. 'That will happen to impressionable young girls.'

'His enthusiasm was catching.'

'Let's hope you're not interested in the fey poetry that's so popular these days. Rambling tripe that artificially excites the senses.'

'Wordsworth, you mean? I'm very fond of his poetry. I don't find it fey, but invigorating.'

'I see.' He glanced at her, but said no more.

They passed through a gate – unlocked and carefully relocked by Arthur – and came into a small pasture. A few placid-looking cattle munched on the grass: impossible to think of them rampaging.

'I may pursue botany now I have come home,' she said.

'I didn't know you had a scientific bent, sister.'

'I don't claim that I do. But the seeds of interest have been planted, I suppose.'

'What area?'

'I am as yet too ignorant to know. I shall start with what is to hand – wildflowers and such – and see where it takes me.' Arthur's legs were long; she lifted her skirts to better keep up, stepping across cowpats. 'Our cousin – Thomas Taylor, you know of him? – studies the cryptogamic plants. I may follow his example.'

'Taylor? I have not met him. Well, there can be no harm in your having a pastime.' He changed the subject to his children, his plans, his tenants. Later though, back in the house with Matilda and Leonora, he said, 'We didn't know our sister had received the beginnings of a scientific education. She intends pursuing botany. I consider it an excellent idea.'

'It seems you have impressed your brother,' Matilda said. 'That's not easily done, I promise you.'

Ellen's eyes met Arthur's. To her surprise they had softened under the gloomy eyebrows; she saw there a spark of kindliness, a sudden light that rescued his face from severity, suggested an acknowledgement of kinship that might be cultivated, with time.

He abruptly left the drawing room, returned after a minute with a book in his hand.

'Take this,' he said, handing it to Ellen. *The Young Gardener's Best Companion, for the Thorough Practical Management of the Kitchen and Fruit Garden*, by Samuel Fullmer. 'In case your botanical interests should lead you to the humble kitchen garden, sorely underdeveloped at Ballylickey, in my opinion. I had hoped that my wife might take an interest here but it seems she will not.'

'My talents lie within the house, not beyond it.' Matilda leaned towards Ellen. 'Each to their own ... my husband is besotted with trees, and fancies I should be also, though he would settle for my becoming queen of cabbages.'

Difficult to imagine Arthur 'besotted' by anything: he gave the opposite impression, of a man who exerted much energy in tempering his passions.

Later, as they said their goodbyes – light seeping from the eastern sky, Leonora refusing to consider a night away from Ballylickey – Matilda implored them to visit again. 'It's been almost six months since you last came, Mother.'

'I venture hardly at all from Ballylickey these days,' Leonora said. 'Bantry, occasionally.'

'You're not ill?' Arthur asked, sharply.

'No, no. But I confess to a weariness deep in my bones that leaves me breathless. Even now, I'm longing for the comforts of my own hearth.' She smiled. 'Old age, I fear.'

'You're not old, Mother.'

'Yes, Arthur, I am. And my body is worn out. But I am well tended to, and so long as I can see the bay from my window, I have a reason to rise in the morning.'

Margaret came first to kiss her grandmother goodbye. Freddie followed, grudging but compliant, holding up his pretty, sour puss. Arthur nodded; Margaret petted his fair head and he ran off. Obviously someone had threatened, or bribed, him to better manners.

The road was no better on the way home.

'Is Arthur ill?' Ellen asked.

Leonora looked at her. 'Why do you ask?'

'He appears … worn.'

'He has many responsibilities. He's stronger than he appears, I promise you. A will like iron.'

Ellen let it go. 'How beautiful the estate is,' she said. 'Or, will be.'

'It was his dream for many years, you know.'

'The house?'

'Land, of his own. He's not the oldest son. He has been bent on establishing himself as a landowner of merit since he was a very young man.'

'And now he's achieved it.'

'Perhaps.' Leonora closed her eyes, as if in acceptance of her discomfort.

'I'll have Kate prepare the hip bath.'

'There's no need.'

'Nevertheless, I think it best.'

'As you say. My remaining teeth feel shaken from my gums.' Ellen took her hand, as if she could absorb some of the jolts. After a minute Leonora said, 'The hunger for land is one of the deepest there is. I find those who harbour it are rarely satisfied.' She shook her head. 'Emanuel is the same.'

'Yes?'

'More so, perhaps.'

'You rarely speak of him. And he writes so seldom.'

'Though far from here, he is part of this place.'

'There's such a gap in our ages. Was he always …?'

'Always? Always what?'

Ellen looked at her hands. 'Angry.' She hesitated, then added, 'Bitter.'

'He did the best he could after your father died, man of the house at such a young age. Then Katherine …' Her shoulders drooped and she looked out the window. Ellen could barely hear her over the noise of the road. 'They were close. Out on the horses every chance they got.' She sighed. 'He changed, after. Less trusting of the world. Then he went to university and somehow found more trouble.' The carriage jerked on, mercilessly. 'My first-born. What need do I have to understand him? He is my own soul. As all my children are. Dead and living.'

∽

The mornings dawned full of promise and flew by, but the evenings dragged. Ellen scoured the house for more books. As she'd feared: nothing but old newspapers, the family bible, prayer books, pamphlets on farming, hunting, sheep breeding, gardening. Someone had buried a pile of political tracts, years old, in a drawer, under a sheaf of old bills. Thomas Paine, *The Rights of Man*; Wolfe Tone, *Argument on Behalf of the Catholics of Ireland*; Thomas Russell, *A Letter to the People of Ireland: On the Present Situation of the Country*. And more. She leafed through them. Words of anger and self-righteous passion. Earnest entreaties, impotent pleas.

A piece of paper fluttered loose, landed on the floor. She picked it up. A fragment of a letter.

Hutchins – I will write with the date when I have notice of it. I may have need of your help, if only to provide a safe bed for

*a night or two, if your excellent mother is agreeable. Memories of
past endeavours reassure me I may rely on your steadfastness. The
signal of light we agreed upon when …*

It ended, torn beneath the last sentence.

She rummaged around the room for a deeper hiding
place, found none suitable.

The kitchen was empty, Kate busy in the dairy. Ellen fed the
smouldering fire with the tracts she'd found. She hesitated over
the note fragment. The words conveyed a powerful undertone of
longing and urgency; it seemed somehow wrong to destroy something so intimate of Emanuel's, presuming he were the recipient,
as he must have been. Yet its existence seemed dangerous, a link to
another time, one of folly and disaster. While she stood there, she
heard a clattering: Kate, stamping clean her shoes at the back door.

She bustled in. 'What are you doing there, Miss?'

Ellen threw in the note, watched it catch flame. 'Some old
papers I wanted rid of.' Kate narrowed her eyes but made no
more comment.

Later, she flicked through a less inflammatory pamphlet
on potato cultivation; the illustrations put her in mind of the
botanical volumes by William Withering that Dr Stokes had
given her. She had placed the books in a closet in her room,
along with the microscope. Now she fetched them out, setting
up the microscope on the table in the parlour.

'That's a nice toy,' Jack said. He played with putting his
thumb under the glass. 'What will you use it for?'

'The study of plants,' Ellen said.

'To what end?'

'Amusement. Learning.'

'That's something you're interested in?'

'Perhaps. They were keen botanists in Dublin.'

'Who are "they"?'

She shrugged. 'Dr Stokes. His friend, Mr Mackay of
Trinity College. Our cousin, Mr Taylor …'

'Ah, yes, our cousin. What kind of man is he?'

'Young,' she said. 'Lighthearted, yet serious, at one and the same time. Energetic. Sharp and clever.'

'Physically?'

'Dark,' she said. 'Not too tall, though well built.'

'Taller than you?'

She blushed. 'Slightly.'

Fortunately he was still playing with the microscope. 'You saw a lot of him?'

'Some. He was generally busy.'

'Staring at plants through one of these?'

She smiled. 'Not just that. Also studying. He's to be a physician.'

'How industrious.' He yawned. 'Any interest in country pursuits?'

'He loves to walk.'

'Hardly a pastime for a gentleman. Still, we should ask him here. He's our cousin, after all. You're already acquainted. No doubt Mother would like to meet him and interrogate him on his family tree. He's amiable?'

'Very.'

'Well, then. Write him that Mr John Hutchins extends a formal invitation to his cousin, Mr ...?

'Thomas.'

'... I Mr Thomas Taylor, to visit Ballylickey House at any time of his convenience.'

She'd started idly leafing through the pages of the Withering. She paused. 'You're serious?'

'Quite.'

'You would have to entertain him, make conversation.'

'I think I could manage that.'

She raised an eyebrow. He looked at her. 'Have you something against him?'

'Why ask that?'

'You painted a glowing picture. Now you seem reluctant to see him.'

'Not at all.' She continued turning the pages of her book, frowning. 'It's just that … I didn't get to speak to him before I left.'

'Well, it's all the one to me.' He handed her back the microscope, pulled himself out of his chair, rested a moment on his crutch. As he passed he peered over her shoulder. '*A Systematic Arrangement of British Plants.*' He leaned over and tapped the cover of the book. 'Are the plants of this peninsula listed?'

'Most, I'm sure. Though Ireland is, geographically speaking at least, another country.'

'Well, if you seek amusement, those volumes will occupy you for months.'

'I hardly know where to begin, though I have read a little on the subject before.'

'Plunge in, I say. I've told you before, we learn best by "doing".' He leaned briefly on her shoulder. 'Write to Mr Taylor. I believe you wish to see him more than you pretend.'

She opened the Withering on a random page: *The specimen of any plant intended for the Herbarium should be carefully collected when dry, and in the height of its flowering, with the different parts as perfect as possible, and in the smaller plants the roots should be taken up. It should then be brought home in a tin box well closed from the air …*

That night she wrote to Dr Stokes.

Dear Sir,

If I could prevail upon your kindness once again and ask you to procure for me a vasculum, similar to your own … Please let me know the cost and I will ensure you're reimbursed.

All goes well here; we are in good health. Remember me, please, to Mrs Stokes and the children, and any of my acquaintances you happen to meet.

Yours,
E. Hutchins

Three weeks later, by parcel to Cork, and from there to Bantry, the vasculum arrived: smart, neatly made, painted sage green to keep off the rust. He'd included a note:

My dear Miss Hutchins,

I am glad to enclose this item as a gift and take your requesting it as proof that you intend putting it to use. I look forward to details of your progress.

Begin with the simplest plant possible and progress carefully, surely, as though finding your way in the dark.

Persevere, with a stout heart, sharp eye and careful fingers.

Your obliging friend etc.,
W. Stokes

Carefully, surely – that was at odds with Jack's advice to 'plunge in'. Perhaps she could do both, as far as possible, by 'plunging carefully' – resolving to commit to the experiment, until such a time as her energy or mental ability might give out, while being as diligent as possible in pursuing the fine ideals set out by Withering. All she risked was pride and time: for the first, who would know or care if she failed? And she had a free supply of the second, once she completed her daily duties.

She had covered basic ground with Rousseau and Mrs Wakefield, but resolved to begin again, as though entirely ignorant. Withering laid it out clearly, logically enough, comparing the divisions within botany to the structure of a country:

The Plant Kingdom to the Kingdom of England:
Classes – Counties;
Orders – Hundreds;
Genera – Parishes;
Species – Villages;
Varieties – Houses.

She preferred his more domestic comparison:

Plants – the Inhabitants in general;
Classes – the Nations;
Orders – the Tribes;
Genera – the Families;
Species – the Individuals;
Varieties – the same individuals in different circumstances.

Increasing the complexity of the system, he added layers of intimacy that broke down and pinioned the very soul of a *flower*, the plant's singular method of reproduction: calyx, blossom, stamens, pistils, seed-vessel, seeds, nectaries, receptacle.

It only took a week's frowning concentration to grasp the fundamentals. Light began to emerge through the fog. Then came the plunge. She searched the hedgerows on her walk and found beside the river a branch of furred, yellow catkins trembling in the spring sunshine. She took out a small, sharp knife she'd found in the shed, sliced a section cleanly away, and carried it home to Ballylickey.

Dissection notes, spring 1805
    Birch – caterpillar-like pods, hanging loosely on a branch – cat-kins, some containing only stamens within their scales, and others

only pistils; in the former, each floret containing four stamens, and in the latter two pistils; Class: Tetrandria; Order: Digynia; four genera, the second and third bearing the male and female flowers in separate catkins; of three possible species, the shape of the leaves indicate Betula alba, or common Birch Tree.

Through spring and into summer she peered into her microscope, poking with her scalpel.

Wood-bine Honeysuckle – intensely sweet smelling, spidery pale yellow flowers on a vine – five stamens in each flower, and the anthers not united, one pistil in each flower; Class: Pentrandia; Order Monogynia; flowers of one petal superior or above germen; seeds in a vessel; four genera, the first three have capsules, but in the last the seed-vessel is a berry with two cells; inequality of blossom; knob at the top of the pistil; identified as Lonicera periclymenum.

Slicing into a flower, she invaded its soul: violently, selfishly, absorbed by greedy curiosity. Folded over the table, squinting in the candlelight, she exposed a dandelion's inner parts, stopped to scribble a note.

Dandelion, piss-a-bed – common, golden-coloured flower, plucked from the front lawn at Ballylickey – compound flower, formed of a number of little flowers (or florets) sitting upon one common receptacle, and enclosed by one common calyx; five stamens with the anthers united and the pistil passing through the cylinder formed by the union of the anthers; Class: Syngenesia; first order; a sort of down adhering to the seeds; scales of the calyx laid one over another like tiles on a roof, the outer scales loose, flexible, and turned back; can only be Leontodon taraxacum.

Like a sharp-edged spark furrowing through her brain, the true nature of the enterprise came upon her. She sat up and stretched, smiling. It was a kind of game, which an assiduous child could play, following clues to the 'treasure' of understanding and identification. And she could bring something

to meet it – a method of recording and setting down her discoveries. She sharpened her pencils, prepared sheets of paper.

A commotion drove her outside: a cart rattling into the
yard, driven by a labourer. Dan followed behind, leading
the pony Jacob as a smaller man might a dog. He called,
'Miss Ellen!'

An inert figure lay in the back of the cart. Ellen shouted,
'Jack!' and ran. As the cart dragged to a halt, she reached
in and tried to touch his foot. Thank God: his eyes blinked
back at her, though a long gash marked his forehead. He
waved her away.

'I'm all right. Don't make a fuss. It was that damnable
Lettie, nipping at Jacob's legs.'

'Have you broken anything?'

'I think not.' His face was almost as white as his shirt collar. 'I'll drown the bitch. I'll tie a stone around her neck and
throw her in the bay.'

'You'll do no such thing. We'll keep her inside from now
on whenever you go out. Where did it happen?'

'In the top field. The gig has a broken axle, damn it
for firewood.'

'Dan, help the master, please,' she said. He came over,
his long face grave. Jacob stood, seemingly unharmed, peace-
ably flicking his tail. As she watched, twisting her hands, Dan
lifted and swung Jack easily off the cart.

'Will you move around for a bit, sir?' he asked.

'No. I'm rather shaken. Take me inside.'

When Dan laid him on the sofa in the parlour, Jack gri-
maced and sucked in a breath. Dan waited a minute, clutching
his cap, conspicuously awkward and looming larger than ever
beside an easy chair.

'Thank you, Dan, I'll manage now,' she said. He bowed and left. She brought cushions and propped them behind her brother's back, took out her handkerchief and dabbed at his forehead. He raised his chin, passive as a child, and let her do it. Close up, the wound was no more than a deep scratch. It had already begun to dry and crust. In the dim light, Jack's expression was slanted, his mouth and chin twisted.

Leonora rushed into the room. 'Kate says that Jack's been thrown from the gig ... oh!' At the sight of him, she stood stock still, her eyes widened with fright.

'He will live, Mother, the wound seems only surface.'

'Should we send Dan to Bandon to fetch the physician?'

'No need for that,' Jack said. He waved his hand towards the dresser. 'Bring me the accounts, Ellen, paper and ink.'

'Rest a while, at least!'

'I must look through this month's bills.'

'You've had a fright.'

'I'm winded, that's all. Nothing to stop me getting on with other work.'

'Have you any headache?' Leonora asked.

'I suppose a headache may start, from too much fussing and clucking.'

'What happened?'

Jack didn't reply. Ellen said, 'Lettie startled Jacob again. The gig needs repairing.'

'That dog is nothing but trouble,' Jack said. 'You must keep her tied up, or I won't answer for my actions.'

Leonora sat down. 'It's dull for her in the house all day. Terriers are curious animals, high-spirited.'

'We'll manage the dog, somehow,' Ellen said. 'Are you cold, Jack? Will I have the fire lit?'

'We've finished with fires during the day. We need to pre-serve wood.'

'At least let me get you a clean jacket.' His sleeve had a long tear and dirt scuffed the side.

'I said, stop fussing.'

'You can't wear that one. And I must wash my hands, I have blood on them.'

Upstairs in Jack's room, Kate wrestled with a bed sheet. She spotted Ellen in the doorway and held it aloft. 'Wine stains, Miss, and God knows what else. I'll have to soak it in lye.'

'He must be nodding off still holding his glass,' Ellen said. 'I'll speak to him.'

'How is the master?'

'His pride is dented. Can you find him another jacket? The other will need cleaning and stitching.'

'He only has his Sunday one.'

'He's not using it for church. It will suffice for now.' He'd refused to attend the last couple of months.

'Very well, Miss. When I'm done here.' Kate cast a fresh sheet across the bed; as it billowed and fell, it floated a pure, cold scent into the room. Outside air, a sea breeze.

'I need to wash my hands.'

Once she went back downstairs, Jack would need cajoling out of despondency, her mother would want her to read aloud for the hours before dinner. Leaving scant time before bed. She closed the door of her room, went to a corner and prised open a loose floorboard, retrieved a small key. Taking the key over to the dresser, she unlocked the mahogany box where she stored her correspondence. She sat in the chair beside the window and reread Caroline's letter.

Prosperous, Co. Kildare
June 1805

My dear Ellen,

After reading your letter, I was agitated for days. I thought myself cut off from civilisation, but in comparison to your situation I am fortunate indeed. Neither a comfortable house nor the goodness of relatives can compensate for regular opportunities to visit friends.

I am quite well, though restless; I have tired of the walks around here and cannot settle at either embroidery or crochet. We redecorated the dining room, a striking pale blue and gold scheme with silk paper, and new plasterwork by Italians brought from Dublin. They toiled from early morn to late at night, working by candlelight. The maestro himself was old, fifty or so, but his apprentices were much younger and exotic looking, with black eyes and hair and square, white teeth. They swung about with great ease and skill, causing much excitement amongst the servants who spent their days inventing reasons to come upstairs and peer at them. I admit to enjoying watching them myself. It's remarkable how such coarse hands are capable of such delicate work – putti, vines and roses like spun sugar, utterly lifelike. I have a stiff neck from gazing upwards. Now they've gone there's a flat feeling in the house, like the day after Christmas.

At least the room looks extremely elegant. I'm planning a dinner to show it off. Not that any of our neighbours would appreciate a modern interior, but there are gentry farther afield who know style.

However, I've been sick recently in the mornings, and I fear I may be in a <u>certain condition</u>. So if I am to entertain I must do so before I am too big to fit into any of my *good dresses. Dear Ellen, I suppose I must be glad, though had I a choice, I would rather not endure it. Who else can I confess this to – who else loves me enough to excuse my selfishness?*

I think you are wise to take Dr Stokes's advice. I have no interest in or time for such activities, nor do I understand their importance, but if it gives you a reason to remain cheerful and forget your troubles, then I am glad. They say botany is still fashionable, that many ladies from the best families engage in it. Remember how entranced we were by Madame Praval's book of Mrs Delany's collages? Could you not do something similar? I should like some to hang on the walls in my little chamber.

Your relatives are fortunate to have you with them. You must make a pleasing diversion. How wild you make the countryside around you sound!

Mr Aylmer has just come into the room and I must finish. Write me soon, if only a note to say how you do.

Your constant friend,
Caroline Aylmer

The troubles they'd cradled as girls now seemed insignificant, ephemeral as snowflakes. Ellen folded the letter back in the box, shut and locked the lid. She tucked the key away in its hiding place. Her mother's voice quavered upwards, from the direction of the stairs. 'Ellen, where are you? Your brother wants you.'

Jack. She stood and called back, 'I'm coming.'

Later that night, before bed, she noticed a sliver of light under his door and went to knock, wanting to have a word with him about the stained sheets. Kate had enough work, he should take more care. Just in time, she heard a curious sound, and paused, lowered her hand.

A low, horrible keening, muffled as if from under the bed-clothes. The sound of secret sobbing, that she recognised from her schooldays, though rarely as primal as this. She couldn't bear it, stepped away, praying the boards wouldn't creak, that he hadn't seen her shadow beneath the door.

The next day, he seemed recovered, quite his insouciant, way-ward, self. He asked for the small back parlour to be cleared. 'Whatever for?' Ellen asked.

'In truth, I find I can manage the stairs no longer,' he said.

It took courage to admit it. 'Certainly, if you want,' she said. 'We can make it comfortable. Your bed, though, will have to be taken apart before moving it. I'll have to send for a carpenter.'

'A simple cot will suit me just as well,' he said.

'You can at least have your own bed,' she said, 'that you've slept in since you were a boy.'

'Aye,' he said. 'Though I wish I might have moved on from it before this. It's likely now I'll die in it.'

'Such dark thoughts so early in the day,' she said, lightly. 'We should consider ourselves lucky to die peacefully in our beds.'

'A sign of a life poorly lived, in my opinion. If I could choose when my end might come, it would be while engaged in some adventure, far from here.'

'And what would we do without you?'

'Already you do almost as much book-keeping as I.'

'Only to ease your load,' she said. 'I have not your affinity for it.'

'Yet you do it well,' he said. 'As you do most things.' He swirled around the dregs of his tea, swallowed it down. 'Despite your affliction.'

'My affliction?'

'You're a woman. While not crippled, in some ways you're as limited as I am.'

She changed the subject. 'You would never leave here, you love this house. And the lands.'

'That's as maybe. The world is wide and varied; I'm sure I'd have found another corner of it equally amenable.'

'What might you have done, had you not had your accident?'

'Gone into the army, maybe. Not for patriotic zeal – I've none of that, damn England, and Ireland besides – but for the opportunity to travel.'

'Mother would not have wanted that.'

'Blast it, there's more to life than what Mother might or might not want. Dear as she is, she has us bound to her side, without hope of escape.'

'There's little point in such thoughts.'

'It's natural, even proper, for a woman to stay at home,' he said. 'Less so for a man. I should have followed my brothers and established a household of my own.'

'You don't think I might want the same?'

'No need to raise your voice, sister. I only mean society looks with less pity on a daughter who stays at home, than a grown son.'

'Society seems to me a capricious judge, of people who generally do the best they can,' she said.

'You should avoid offending it. But nor should you worry unduly about earning its favour. That, at least, is an advantage of living away from it.'

Withering advised the *procurement of the following apparatus*: *A strong oak box of the size and shape of those used for the packing up of tin plates; A quantity of fine dry and searced sand of any kind, sufficient to fill the box; A considerable number of pieces of pliant paper, from one to four inches square ...*

Some flat leaden weights, and a few bound books.

The oak box she had made by a carpenter in Bantry; she ordered weights from an ironmonger in Cork. Dan carried buckets of sand from the beach and she sifted out any pebbles with a wheat sieve, before spreading quantities of it in the oak box to dry before the fire.

'How does this apparatus work?' Jack asked, lifting the lid. Ellen sat cutting up squares of paper at the parlour table, close to the window to make the most of the last light.

'The plant is laid on a sheet of paper, like so.' She took a bluebell from a jar Kate had left on the table. 'Then I cover its parts with a system of weights and sand. Finally, I set the box before the fire, or in the oven. In two to three day the plants will be perfectly dry.'

Too heavy-handed an application of weights and sand, too long left before the fire, resulted in shrivelled, discoloured scraps that could only be thrown in the flames. Too light a touch meant more robust plants simply sprang back up at her when she removed the sand and paper, unaltered. Hard, to get the right balance. It took hours, in wind and rain, to collect samples; if the process failed she must venture out again, hope to find fresh samples in the same spot. She tried again, failed, tried again, learned to engineer better luck. Eventually she became used to the degree of pressure required so as not to squeeze out the delicate juices that gave the plant its colour, the positioning of the weights, the exact length of time to leave the box on the hearth. Kate made no comment when she found it on the shelf above the cooling ashes. Strange pastimes of the young mistress.

SEVEN

Having sent word some weeks previously, James Townsend Mackay arrived on a damp July morning after breakfast. As he came into the parlour, his trim form dusted with drizzle, she jumped to her feet.

'You see I have not forgotten you, Miss Hutchins,' he said. Kate hovered near the door. The last stranger to call had been a notary from Cork to see Jack, in February.

'May I introduce you to my mother?' Ellen said. 'Mother, Mr Mackay has recently been appointed Curator of the Botanic Gardens of Trinity College, Dublin.'

'Delighted, Mr Mackay,' Leonora said. 'Will you take tea?'

'I won't at present, thank you Mrs Hutchins,' – Mackay peered through the window – 'I see the mist is lifting. My limbs are stiff from the ride. Perhaps your daughter could accompany me for some air. Miss Hutchins? Have you time to spare?'

She did.

While walking, he gave her news of friends. He had gone 'herborising', as he called it, with Dr Stokes the previous month, in the woods at Luttrellstown; later at Harcourt Street, William had been brought down from the nursery to be shown off. He was, according to Mr Mackay, a startlingly alert child who at nine months could almost walk and already spoke some words. 'And of course I see your cousin regularly.'

'How is Mr Taylor?'

'Very well. A trifle fatigued mentally following the long academic year. But in excellent spirits.' He paused. 'It occurs to me that you are rather alike,' he said.

'Mr Taylor and I?' she said. 'Surely not.'

'Not in looks, I grant you,' he said. 'Yet there's something.' His r's trilled and rolled like music. 'Perfectly natural. You are cousins, I'm told.'

'Removed.'

'Indeed,' he said. 'Ah, the rocks!'

'There's a shingle beach below.' The sun broke through, briefly, and slipped away again behind a cloud; the Irish summer, a perpetual measuring of light against dark, as if neither should dominate. They climbed down to the beach. With his cane Mr Mackay lifted the edge of a mass of brown seaweed.

'Bladder rack,' he said. 'Common on the Atlantic coast. How goes your botanising? Dr Stokes tells me you've made a start.'

'I have no one to ask for help or instruction. I long for a like-minded enthusiast to talk with. Alas, the only such friends I have are hundreds of miles away.'

'Your family support your endeavours?'

'My mother knows how important it is for me to occupy my mind with some useful pursuit.'

'Fortunately you have unique access to this treasure trove.' Waving an arm to embrace the bay, the mountains. Then he prodded again at the thick seaweed. 'Have you considered making algae your speciality?'

Ellen bent and plucked a strand. 'Is it so interesting?' Its flaccid, finger-like form made it appear scarcely a plant at all.

'Firstly, algae are vastly unexplored. You'll find little in Withering but the briefest introduction. Secondly – do you not think the form of the plant itself, hidden, as it often is, underwater, along with the strange fructification, makes for a fascinating subject?'

She turned the strand over in her hand. 'I can hardly tell the root, stem and leaf apart.'

'We distinguish the genera by the situation of what we presume are the flowers or seeds, or by comparison with similarly formed plants we're more familiar with.' He pointed to the bulging pods, like green pearls within the clump. 'These bladders carry the seeds, male and female. Let us take a sample now and study it later.'

'Will you stay with us long?'

'A few days, if I may. I'm interested in the climate here, how it impacts on the plant life.'

'We scarcely ever get frost.'

'A singular corner of the world.' He turned from the sea and faced her, a swift motion that made his coat tails whip against his legs. 'As different from the hurly burly of Dublin as could possibly be.'

'I am getting used to silence,' she said, carefully. 'My mother's house has few visitors.' She wished he would not look at her with so kindly an expression! Sympathy was ever likely to make her falter.

'Do you often see acquaintances your own age?'

'Apart from my brother, not at all.'

'I see.' He cleared his throat. 'You know, Miss Hutchins, I have often found in my own life, that what initially seemed an unpalatable situation revealed itself in the fullness of time as an opportunity. When I came to Dublin as a friendless young man, I thought I had made a mistake. I sat alone in my accommodation by the quays, staring out the window at the chaotic scenes below, wondering if I should leap aboard the first available boat and return to Scotland.' His eyes darkened, flicked out to sea. Then he smiled. 'But in time I adjusted, with the help of my benefactor Dr Scott; once I travelled to the west of Ireland and began my studies in earnest, better fortune presented itself.'

'I'm sure what you say is true, sir.'

'I hope you don't mind my enquiring as to your frame of mind.'

'You may do so unreservedly.'

'I shall report back to Dr Stokes that you appear content, at least.' He held out his palm, squinted into the sky. 'Shall we make our way back? I shouldn't like for you to get a chill.'

'The drizzle causes no great harm.'

'Spoken like a born naturalist.'

They came to the boathouse. Mackay indicated the single boat inside, of good size but in a melancholy state of disrepair. 'Does this belong to your family?' he asked.

'Yes. Though unused for many years.'

He walked ahead and proceeded to examine the vessel. 'She was a fair craft, once.'

'You're familiar with boats?' she asked.

'I spent time in them as a boy.' After studying the boat from every angle, pulling at loose boards, he said, 'Consider how useful this would be to you, were it sea-worthy. You'll need to visit the islands and the more inaccessible coves, particularly if you're to collect the sea plants.'

She ran her gloved hand over the hull. Flakes of faded blue paint came away. A large gash showed why the boat had been originally laid-up. Mr Mackay dismissed this. 'Some new planks, retarring, repainting – no doubt there are many local men qualified for such work. And you'll need a crew.'

'I'll speak to my brother,' she said.

'I cannot decipher the name, it's quite faded away,' Mr Mackay said. 'You must find it out, or rename her.'

Leonora met them at the door. 'Mr Mackay, you'll take tea now, I'm sure?' she said.

He bowed. 'If you had something stronger in the house, I should prefer it.'

She blinked at him. 'Wine, you mean? Or whiskey? I think we have some whiskey. I cannot vouch where it came from.'

'Legal or illicit, madam, either is acceptable,' he said.

Ellen called for Kate, and drew her aside. 'Bring whiskey for Mr Mackay.'

'He's a fine enough looking gentleman, though he speaks so I can hardly understand him. Now he takes whiskey in the morning like a man from the cottages.'

'We don't question the habits of our friends,' Ellen said. Kate went for the whiskey, shaking her head.

'See here, Ellen,' Leonora was saying. 'Mr Mackay has something for you from Dr Stokes.'

He held out a parcel. 'He presumed it might be difficult to purchase materials in these parts.'

When she opened out the packing paper she exclaimed. Dr Stokes had sent her a paint box, of shining mahogany, with cakes of colour, glass dishes for mixing, a marble palette. She took one of the brushes and swept its soft bristles across the back of her hand.

'My daughter is speechless, Mr Mackay,' Leonora said. 'I thank you for delivering another example of Dr Stokes's kindness.'

'I was glad to do so, madam.'

Three days later he left for County Kerry. As they parted, shaking hands, he looked directly in Ellen's eyes. His rolling accent gave emphasis to his words. 'Remember my advice about the sea plants, Miss Hutchins. I correspond with many botanists in England who would be eager to receive your specimens.'

❧

After Mr Mackay's departure, the days resumed their monotonous routine. She walked along the beach, kicking over seaweed and debris, thinking on his advice. The local people used wrack to enrich their meagre strips of land, carrying great baskets of it from the shoreline. A picturesque scene, that masked their fatigue and poverty; the women in bare feet with their skirts tied up, shawls of ragged cloth around their heads and necks against the sun, the men labouring with forks or bare hands, the fortunate few loading up their donkeys. Ellen had never cast seaweed much thought before now. It had neither beauty nor novelty, until Mr Mackay suggested otherwise. He had presented it as worthy of study, a mystery to resolve. The thought of exciting the interest of other botanists had much appeal, were she to prove fit for the task. She wandered to the edge of the waves, where swathes of the dark, shining fronds tossed forth and back, and thought: *It will be wet work, and cold.*

Mr Mackay's words: *You have unique access to this treasure trove.* The boat. A pair of stout boots. Her knife. She would try.

The shipwright arrived two weeks later with a team of ten stout men. As they lifted the boat, it strained and creaked as if ready to fall asunder. At last it was hoisted onto a wheeled platform and a pair of farm horses dragged it off to the yard in Bantry.

In early October the weather became fine, ripening into golden days and brief, purple-tinged evenings. The shipwright sent word that the boat was sea-worthy and ready to be sailed around the bay, back to Ballylickey.

A procession set out from the house. At its head, Ellen: next, Jack, clinging to Dan's back. Then Kate, carrying a chair. They stopped at the water's edge and Dan lowered Jack down, where he sat like a Venetian awaiting a flotilla, shielding his eyes against the glare of the sunlight on the water.

At last a small, dark shape came into view. As it came closer, the crew of two became discernible: bare heads, stout arms pulling in harmony. Easing off the oars, they guided the boat into the cove. Shining with a new coat of paint, it was scarcely recognisable as the same vessel.

'Ho, there, Mr Barry,' Jack called. The older man of the two raised a hand. Standing, balancing expertly, he jumped onto the rocks.

'How does she go?'

'Water-tight and sweet as a nut, Mr Hutchins, sir.' Mr Barry jerked his head towards the boat. 'Are you coming out?'

'Not I,' Jack said. 'As I told you, my sister will command this vessel.'

Mr Barry had a grizzled, impenetrable face, leathered from sun and wind. 'Very well, sir.'

'Take your direction from her, the boat is entirely at her disposal.'

'Will we be taking the boat far, sir?'

'Ellen?'

She stepped forward. 'Good afternoon, Mr Barry. Not so far. Garnish and Whiddy islands and all in between.'

His dark eyes swivelled towards her, unblinking as a lizard's. 'As you say, Miss.'

'My sister has mapped her course,' Jack said. 'Ellen, will you try her out?' The boat rocked and creaked against the rocks; beyond, the waters of the bay stretched wide, out to the boundless Atlantic. 'Ellen?' Jack's voice was firm. 'Mr Barry is greatly experienced. I've known him many years.'

'The lad, Michael, is my own son,' the seaman said. 'Come ahead, Miss.' He extended his hand, as though guiding her to a dance. She came forward, put her hand in his, allowed him help her onto the boat. It shifted and rolled beneath her feet. 'Steady, Miss, find your balance.' She nodded at the younger man and stepped awkwardly to her seat, a bare plank set across the hull.

Mr Barry climbed aboard in two easy strides and spoke to the young man in Irish. In smooth unison they pulled on the oars, manoeuvring out of the cove. Ellen waved at those left on the rocks. Back, forth, back, forth. The land began to recede. Here was an unfamiliar view, a shift in reality. Beyond the cove, the rocks, overrun with little waterfalls. Shingle, headland, Ballylickey House amidst the trees. With each stroke a sensation of separation, a lightening. Expansion. She eased her grip on the edge of her seat. The men talked amongst themselves in their own language, one she could not penetrate.

She must grow accustomed, somehow, to giving orders to these two men, with their impassive faces like polished wood. The *Ballylickey Lady* was hers to command.

Excitement soon gave way to frustration: two days later the weather broke and autumn raged in with a series of storms that made even her daily walks impossible. The boat was laid up again, to wait for spring.

One morning, while the wind rattled the window-glass, she set to re-ordering her dresser drawers; at the back, pushed beneath her underclothes, she found a small, thin packet.

She kneeled on the rug and unfolded the yellowed, papery flap of skin, dried to translucence: William Stokes's caul. She refolded it, put it back under her clean shifts, and went on with her tidying. When finished she shut the last drawer, stood to go.

Her legs wouldn't move. She leaned against the dresser.

Before she could think better of it, she opened the drawer where she'd buried the caul and snatched it out again. She went to the fire and threw it on the smouldering logs. It flared and burned into wisps of sparks that vanished up the chimney.

Prosperous, Co. Kildare
May 1806

Dear Ellen,

I'm enclosing one of Charlotte's curls. As you can see, it's much the same colour as my own. She is a bonny four-month-old now and eats, drinks and laughs as good as an angel.

For the first weeks after giving birth I felt very anxious and irritable, and gave in to crying fits and all manner of childish behaviour ... Either that or I felt numb, as though I'd been fed a strange elixir. I was fortunate to have a good, clean wet nurse, a local woman, for much as I told myself I should love Charlotte, I found it difficult to do so. Worse still, there were times I wished not to have anything to do with her, and could have given her away to the first beggar woman that came knocking at the door.

No doubt you think this dreadful, but motherhood has been like being turned inside out. Even now, when I'm so much better and able to enjoy Charlotte for the darling she is, I think again of my own mother and could collapse in a fit of weeping at any moment. Do you remember how foolish I was, that night when I crept into your bed? I believe, as you said I would, that I will one day feel less like a piece of muslin, to be blown here and there without direction or influence. Charlotte may yet help.

Mr Aylmer is very tolerant and truly fond of me. He doesn't ask for more than I can give. From what I'm told, this is an admirable quality in a husband. He is, moreover, exceedingly gentle, I believe he would not strike even a dog. I hope you're settling in to your new life. You were ever more surefooted than I.

Affectionately,
Caroline Aylmer

YARMOUTH, ENGLAND
JULY 1806

EIGHT

Ink dripped from the desk onto the floor; a merry, ticking sound. Dawson Turner flapped at the rapidly spreading puddle with his handkerchief. A young family was a mixed blessing for a man of many interests and not enough hours in the day to pursue them. Instantly, he chided himself. In light of recent events, this was unworthy, verging on contemptible.

His youngest daughter, three-year-old Anne, watched him, sucking on her thumb. Her clear eyes seemed to say, 'What will you do now, Papa?'

'Rachel! For heaven's sake, Rachel!'

She came running. 'Oh, sir. Did you do this, Anne?'

'Leave it,' Dawson said. 'I'll see what can be salvaged. Take the child and keep her out of the study.'

'I'm sorry, sir. I don't know how she got down here. I only turned my back for a minute. Everything upstairs is that topsy-turvy.'

'I know.' He rescued a sheet of paper covered in his sprawling handwriting and held it, dripping, aloft. 'Things will settle in time, God willing.'

'If you say so, sir.'

To his surprise, she seemed close to tears. 'There, now, no serious harm done,' he said quickly. 'Only notes, nothing very important.' Rachel, usually so unflinching in the face of all manner of household disasters.

'I'm very sorry, sir.'

'Nothing that can't be wiped away.' Though he feared the stains might linger, evidence of the incident for years to come. 'Where is your mistress?'

'In her room, sir.'

'Still?'

'Yes, sir.'

'I'll go to her shortly.'

Scolding and cooing, Rachel carried Anne away. He spent ten minutes saving what papers he could, and laid them on the hearth to dry. Another servant answered his bell; he ordered her to clean the stains from the desk and the floor. As she set about sponging and rinsing, he got to the pleasurable task he'd left until the end of the day. The latest parcel from Dublin had arrived while he worked downstairs in the bank; he'd only had time to sign for it and order it to be taken upstairs to his study. Fortunately, he hadn't left the parcel on his desk, but on a side table with the rest of his post.

As always, Mackay had packed the samples in layers of packing paper and soft muslin. Dawson slowly peeled these back and revealed the dried plants. Carrying each one by one to the window, he held them to the light, examining ribs, fronds and fruit. *Fucus laceratus. Fucus wiggii. Fucus palmetto.* Excellent specimens, beautifully preserved within squares of folded paper. When the desk had sufficiently dried, he'd start work; the book expanded daily, monumental in time and effort …

'Is there anything else, sir?'

'What?' He looked up, torn from his thoughts.

The young servant stood by the door, holding a pail and mop. 'I've cleaned up the spill, sir. As best I can, anyway. Do you want me for anything else?'

'No, that's all. Close the door on your way out.' When she'd left, he picked up the note from Mackay.

Botanic Gardens, Dublin

Sir,

I trust these samples will make it intact to Norwich. The Cryptopleura ramosa *is particularly interesting, as I believe it is seldom found in your part of England.*

I have asked a lady of my acquaintance, a Miss Hutchins, to furnish you with samples for your History of Fuci. *She lives in a unique part of the south of Ireland, around Bantry Bay, and has access to a diverse range of marine plants. She is, besides, a talented plant hunter, capable of finding almost anything I suggest. Regarding botany, her aptitude for learning is only surpassed by her indefatigability. I am sure she will oblige me if I ask her to provide more. It is a difficult part of the country to access and remains quite wild, and is all the more interesting for it.*

I hope the Fuci *progresses well. I admire the energy and ambition that fuels it and hope to provide any assistance I can.*

I remain, your obliging servant,
J T Mackay

If he were honest – and Dawson Turner was almost always scrupulously so, with himself and everyone else (a fortunate attribute for a banker, at least) – part of the reason for his manic enthusiasms, the relentless nights of writing, the gnawing ambition that launched him into vast projects such as the *Fuci* – lay in his disappointment at never finishing his theology degree. His father, owner of a Yarmouth bank and almost as strong and hearty as Dawson himself, fell from a ladder in an orchard, where he'd been sampling cherries. A fit of some kind. Though he initially rallied, his health began to decline from that point on; in Dawson's final year he took to his bed in the grip of a final malaise. Dawson had no choice but to leave Cambridge early and return home. When his father eventually died, the course of Dawson's life had been set as gentleman, banker, husband and father. His fate: to live in the bank he'd inherited, setting up his family on the top floors, conducting business on the ground floor, in the handsome building on Hall Quay facing the River Yare.

It wasn't what he would have chosen. Still, a good life. One that enabled him to fill every free second with *crypto-gamia*.

Nine o'clock: he'd spent almost two hours examining and making notes on the sea plants sent by Mackay. The light of the midsummer evening began to fade. Before he lit the lamps and started into *Fuci* – once he began he might continue working until after midnight without lifting his head from the page – he'd check on Mary. A quick glance into the drawing room. She wasn't there. On the next floor up he pushed open the door of the room she used for her studio. The beginnings of a portrait of a young woman rested on the easel. The sitter had a forlorn expression, as if wondering if she might ever be finished. A sheaf of loose sheets lay on the table – portraits of a child, from newborn to toddler to little boy, along with those of his sisters and others. Dawson lifted the top drawing. How skilfully Mary had captured the boy's lively intelligence and good-natured impudence, his animated expression. He flicked on through the rest. Here was little Dawson playing with a kitten; here he was laughing, crawling in a long gown. And here he was asleep, eyelashes curled on his cheeks. Staring at the dimpled features of his son, a lump lodged in his throat and his eyes smarted.

Mary had not been in here since, though this part of the building had been unaffected by the fire, the damage contained to the attics and nursery. Dust lay on every surface;

cakes of paint had been left open to the sunlight, the pigment drying out. How would she react when next she came in to the room, confronted by these drawings? He picked up one of her pencils and tapped it against his chin. After a moment he threw it down and gathered the drawings in his arms.

Half way out the door, he paused.

She might well be overcome, relapse into hysterics. Or she might find solace in these, the only record of her son's features. He put them back where he found them, retreated, closed the door. Her ghosts: to keep or destroy, as she wanted.

She wasn't in their bedroom.

He met Rachel on the stairs with an armful of linen. 'The mistress is not in her room.'

'She just went out, sir.'

'Out?' Mary never went out alone after dinner, even during the fine summer evenings.

'Across to the quay, sir. She said she needed air. She asked that you not be disturbed.'

'She didn't go out the front door?' This, the main door into the bank, had been locked and bolted as usual since close of business.

'No, sir. She went by the back.'

Pulling on his coat, Dawson hurried through the back door into the yard. From there, once through the gate, he hurried through the 'row', an exceptionally narrow medieval passageway that cut through the terrace and led out onto the quays.

She hadn't gone far. She stood at the water's edge, looking into the deepening sky. She wore no bonnet. He walked as quickly as he could without running, glancing up and down the quays. All quiet at this hour, though it could not be called safe. Rough, sea-faring men from the many docked ships passed along here on their way to the inns and hostelries of the town, staggering back again in the early hours.

Seagulls screeched over the water, seeking out remains of fish. Across the Yare, a group of men made the most of the last light, stacking boards in the timber yard. The sound of wood hitting wood smacked across the river as a cracking echo.

'Mary.' He kept his voice low. She turned. Her lovely face was white. Dark bruises circled her eyes. At night she tossed and moaned in her sleep, sending him to the sofa.

'It's late,' he said. 'Come inside.'

She looked away again at the opposite bank. 'I had to get out. I couldn't breathe. Have you noticed the smell is back upstairs? But worse than it was, as though something rotted under the floorboards?'

After the fire there had been a smell that lingered for weeks, of ash and soot, and a vile under-note, like spoiled tallow. They left the windows open day and night to let in the sea air; he had the walls scoured, all curtains, linen, and furnishings washed, great bunches of rosemary and lavender scattered throughout the rooms. Eventually the atmosphere cleared.

'Something drifting across from the docks, perhaps.' He put his hand on her elbow. 'Dearest, come indoors.'

'I find the water soothing. Painting it, I would have to use pure silver.'

'To my eye, it is more like pewter.' Dawson calculated what to say to make her come back with him.

'Rachel said you had a parcel?' To his relief, her voice was almost normal.

'Yes. More algae from Dublin. Quite unharmed, in perfect condition. Remarkable specimens. It seems Mackay has found another source for sea plants. A young lady from Bantry Bay, in the south of Ireland.'

'A young lady? Unusual.'

'It shouldn't surprise you. I know of a Mrs Griffiths, for example, an exemplary plant collector.'

'Still, the young lady from Bantry Bay should be encouraged. You'll write to her?'

'I may do, yes. It would save Mackay the trouble of acting as go-between.'

'Life goes on.'

'Yes.' He moved his hand to her shoulder. 'Somehow, it does.' Did she blame him for that? He led her away from the water. She lifted her shawl and drew it over her hair, leaned towards him.

'Sit with me for a while tonight, Dawson.'

'Of course, my dear.' *Fuci* would have to wait.

BALLYLICKEY, CO. CORK
JANUARY 1807

NINE

She asked Dan if there would be any objection to using the small field close to the house for a garden.

'I don't know, Miss,' Dan said. 'We give the fowl the run of it.'

'And you can still, if we keep them away from the seeds.'

'Best to check with the master.'

Jack had no objection, though he asked, what type of garden?

'Salads, herbs,' she said. 'Also ornamental plants, shrubs, flowers. I'd like to experiment. With the climate here I hope I'll have luck as Arthur has had at Ardnagashel.' Winter had again been exceptionally mild; linen had been hung outside to dry on Christmas Day, a marvel.

'I wonder at your having time,' he said, 'or energy.' He rubbed at his legs. 'While I must sit here, no help to you at all.'

'You still run the farm,' she said. His face had slumped into a scowl. Best not to indulge him, but smile, say cheerfully, 'Wouldn't it be pleasant to have fresh greens by summer?'

He agreed it would. In the end, he shrugged, 'By all means, sister, make your garden. Kate will be delighted to hear you've found another way of soiling your dresses.'

She drew a plan, organising beds for different planting, interspersing rows of edible plants with perennial flowers and shrubs. Mr Mackay, with whom she now regularly swapped parcels and letters, sent seeds and promised to send bare root roses. She plotted and dreamed of a pleasant space close to the house where she could spend hours of leisure, a haven of scent and colour, roses, lavender, herbs of all kinds – sorrel, lovage, sage, rosemary.

'That's some time away, Miss,' Dan said. 'A garden takes many years.' He wiped sweat from his face with a soiled square of fabric, then squirreled it away in his jacket pocket. She'd agreed with Jack not to take the servants away from their work, but every time she set foot in the field, Dan appeared with a spade over his shoulder.

'I know it,' she said. 'But now I've made a start, it will be something to aspire to. Time is something I have.' He glanced over at her, frowned, and bent his back again into the digging.

Arthur visited and was admiring, 'You've made great progress. A glasshouse is what you need next. Start your seeds earlier, extend the season. It need not be elaborate, a basic wooden frame. And, of course, the glass.'

'Would that not be costly?'

'It can be managed. I'd rather you had some shelter when working, you seem to spend an inordinate amount of time outdoors.'

'I like it.'

'Still, you should take care not to catch chills.' He took out a large snowy white handkerchief, coughed discreetly into it. Ellen wondered if a preoccupation with health were not partly the cause of Arthur's malaise, which appeared vague in nature – a general fatigue, a sapping dispiritedness.

'You look well,' she said, encouragingly.

'Oh?' He smiled, sniffed. 'Well, it's one of my good days, I admit. My doctor wishes me to take the water at Harrogate. If I can find time.'

A fortnight later a cart arrived; two men descended and announced they were there on Mr Arthur Hutchins's orders. They unloaded lengths of wood, tenderly packed sheets of costly glass. Once the skeletal frame had been erected, it took them but a day to transform the airy box into a snug, transparent cocoon that glittered in the sun. Cunning windows could be opened to control the flow of air and ensure she could work without becoming too uncomfortable in the heady warmth.

Ballylickey, April 10th 1807

Arthur,

I am quite delighted with the glasshouse. My new plants advance at an astonishing rate. It has become home for an extraordinary number of beetles and other creatures, which also thrive in the jungle-like atmosphere.

We find it impossible to keep the yard cats out, they so love the heat.

Affectionately yours,
E Hutchins

The seeds sent by Mackay sprouted to trefoils, growing daily under her care.

She might have once described herself as placid, or passive; this had been only a kind of suspension, a preserving of energy. When necessary she could be single minded, as the green spikes that poked through the soil after months in the dark. The gentle days of late spring arrived. Evening stretched forward, dark retreated. The house warmed up: not so difficult now to get out of bed, squat over the chamber pot and stand at the basin sluicing face, body, legs and feet with a wet cloth, hurry to dress in front of the window with the view to the sea. The garden began to be tamed, came to life. New smells, green, cool, unutterably fresh, as moisture pulsed off the new grass and unfolding leaves.

Though postmarked April, Dawson Turner's first letter arrived in May.

He conveyed his obligations, thanked her for the specimens she'd sent him through James Townsend Mackay,
offered to forward her any plants she might find interesting
from his part of England. The address given was Yarmouth.
Somewhere on the southeast coast, Jack thought – she pored
over the framed map of England hung in the downstairs hall
and found it at last, a dot on an easterly hump of the English
coastline, jutting into the North Sea. A fishing town, in all
likelihood, like Bantry, though presumably larger. The tone of
Mr Turner's short letter was careful, cordial, as was appropriate from one stranger to another. He enclosed a parcel of sea
plants he thought she might find interesting.

Mr Mackay had mentioned Dawson Turner frequently:
he was an intimate friend of Professor Scott, Mr Mackay's
patron, an avid and exceptionally industrious botanist, a published writer and fellow of the Royal Society, the Linnaean
Society, the Antiquarian Society. Mr Mackay made brief ref-
erence to Mr Turner's personal circumstances: comfortably
well off, if compelled to manage his father's bank, a correspondent of the most eminent botanists of the day, as well as
other famous and noteworthy contributors to British science
and letters. In February Ellen had sat up night after night,
reading his *Synopsis of British Fuci* until her eyes throbbed.

Yet she didn't respond to his letter until almost the end
of June. Delaying the task delayed the satisfaction. And she
needed to study the plants he'd sent her, search out something
of equal interest to write about.

As summer bloomed, the garden and lands consumed her.
Her lettuces, beans and fruits needed constant watering. She
carried the wooden buckets slowly from the pump in the yard,
sloshing water along the way. By the last light of evening she
searched out slugs, snails and caterpillars and drowned them
in a pan of seawater. Jack went out less and less, complaining

of pains in his legs and back, and all work on the house and lands required supervision, if only once a day; she learned to drive his cart, as he seemed to have abandoned it. Bumping over rutted tracks, she travelled the boundary led by Jacob, who knew the way with his eyes closed and needed little coaxing. The fields were spread out, not all easily accessed. Often she had to abandon the cart and walk with Lettie at her heels, the dog barking non-stop, maddened by flies.

Initially the tenants met her attempts at discussion with silence, shuffling their feet and casting truculent glances at her; only for her being a Hutchins, Mr Hutchins's sister, she suspected they would have refused to deal with her at all. In the end she wore them down, turning up repeatedly, smiling in the face of their limp hostility, making her voice authoritative, using the weather as a point of common interest: mare's tails, the skim of ragged cloud like scraps of lace, heralding rain; the swoop of swallows offering a chance to plough. On her way back through the fields she looked for plant specimens. Tomorrow, she told herself, tomorrow I will set out in earnest: tomorrow slipped away like water down a drain; she was forced to wait. To fit in plant collecting, she began to rise at five, sat up past midnight. A fierce energy had overcome her; so unrelenting, she feared it might consume her entirely.

Arthur and Matilda arrived early one Wednesday morning with Freddie and Margaret. The children scrambled down from the carriage and romped towards the yard. Arthur circled the lawn, hands clasped behind his back, while Matilda and Ellen watched over the unloading of the children's trunks.

'Arthur looks tired,' Ellen said.

'He's been under a lot of strain. The physician prescribed rest. Hopefully the waters will help. Thank you for taking the children.'

Ellen took Matilda's arm. 'What troubles him?'

'The estate, for one thing. Not Ardnagashel – his schemes progress to his satisfaction – but your father's.'

'After all this time?'

'Matters yet unresolved, he says.' Before Ellen could reply, Margaret ran up, panting. Freddie followed, hanging behind.

'Aunt Ellen, will you take us out on the boat?' Margaret said.

'Perhaps. If you do as you're told and say your prayers nicely.'

'You mustn't put your aunt to any trouble while we're gone,' Matilda said. 'I expect you both to behave. Freddie, what is that face?'

'I want to go with you.'

She caressed his hair. 'Really, you'd find it very dull. And Papa needs complete quiet.' Freddie whimpered, watching his mother through his lashes. He drew in a breath and burst into a wail.

'That sound is very like that which the pig makes before being dragged off for slaughter,' Ellen said. Margaret laughed, then put her hand over her mouth.

'Oh dear.' Matilda's face crumpled in distress. 'Don't cry, my pet.' Freddie's face grew purple. 'Mama will bring you home a present, anything you like.'

'He's worked out how to capitalise on your affection,' Ellen said.

'He's only eight, sister Ellen, and has a sensitive nature.'

Ellen would have little time to humour Freddie over the next month. 'Good boy, that's enough,' she said, 'Go find Lettie.' Freddie glared at her, unwilling to give up the assault on his mother's nerves. Just then the dog raced yapping onto the lawn, a timely distraction. The boy's wailing dropped to a snivel. He scrubbed away his tears – only ever on the surface – with his fists. Margaret, obviously judging the storm to be over for the present, took his hand and pulled him away.

'He'll settle once we've gone,' Matilda said.

Ellen cleared her throat. 'What did you mean about "matters yet unresolved"?'

Matilda frowned. 'I don't understand the exact details. Something to do with Emanuel.' She bit her lip. 'It isn't my place to say.'

'A disagreement?'

'Enough to rob Arthur of his sleep.'

'I wish Emanuel—' Ellen stopped.

Matilda leaned closer. 'What, my dear?'

'That he were less remote.'

'Far away, you mean? Yes, London is a long way from here.' Not that. Rather that she could know for certain if his regard for her, Jack and her mother was as solid as theirs for him. She assumed it. But did he show evidence of it? They rarely received letters from him; those that did come were full of terse questions about rent, moneys spent and owed, requests for deeds and documents. Their communications seemed little more than might be expected between farm manager and employer. How he lived, whom he saw, remained a mystery. Adopting the same tone, they kept their letters brief and perfunctory, as though anxious not to trouble him with unnecessary detail or fuss. It didn't seem a natural state of affairs between members of the same family, but what could be done? Neither affection nor intimacy could be forced, nor wished into existence.

'I would like to know him better,' she said. Matilda opened her mouth as if to say more, but Arthur shouted across the lawn, they must go, a week's trying journey lay ahead, first to the city of Cork, then to Dublin, where they would catch the packet to Holyhead, followed by a long and exhausting coach ride to North Yorkshire and Harrogate.

'It will be a wonder,' he finished, 'if I make it there alive. Which, considering the purpose of the trip, would be ironic indeed.'

Matilda put her mouth to Ellen's ear. 'Leave it between the men, sister,' she said. She walked away, her face smoothing into blankness.

The children came running at Arthur's command to kiss their mama goodbye. Later Ellen would bring them to the garden, allow them gather lettuce and radishes for dinner. Even children quietened in a garden, once their fingers touched the soil and they breathed the green growing.

❦

A low-lying mist draped across the peninsula. Jack said it would lift, burn away by midday.

Sitting on the three-legged stool by the back door, Ellen pulled on a pair of his old boots. They fit her snugly, the leather soft and slippery under her fingers. Kate had suggested rubbing them with goose fat to make them better resist the wet. She came now into the hall, wiping her hands on her skirts. 'Let me help you, Miss.'

'I can manage.'

'Maybe. But let me do it.' Ellen gave up the argument. Kate seized the boots and wrangled them on efficiently, if not gently. 'There.' She stepped backwards, frowned. 'Are you tying up your dress, Miss, when you're out in the muck?'

'As best as I can.'

'It's easier to get mud out of a petticoat than a dress.'

'I promise to try to keep my skirts up.'

'Hmm.' Kate sounded sceptical.

'I know it brings extra work on you.'

'Well. For you, Miss, I'm glad to do it. I only wonder at how you get into such a state.'

'It's necessary, I'm afraid.'

'I fear you may do yourself an injury.'

'So far I've been neither attacked, molested, blown from a cliff or drowned, as you predicted I would.'

'No. Though it took you long enough to get over that cold.'

The defiance she'd shown Mr Mackay against the rain had proved foolish. Weeks previous, she'd arrived at the back door, dripping and shivering. At the sight of her, Kate's lips thinned. She took charge, wrapped her in blankets, boiled water for a mustard bath, packed her off to bed. Despite these attentions Ellen still came down with fever, after which she lay spent and immobile for days. Kate's face, an impassive moon, floated in and out of her consciousness, a wavering, comforting presence, her rough palm cool on Ellen's burning forehead. She was dimly aware of Leonora creeping in to sit by her bedside and murmur from her bible. Slipping between dreaming and waking, Ellen imagined herself back in the nursery, her mother younger, stronger, capable of nurturing a child back to life.

By now she'd regained most of the weight she'd lost, but Kate still watched her, eyeing any scrap she left on her plate, muttering as she hung the wet stockings before the fire. The illness had confirmed in her mind the madness of what she called Ellen's 'tramping'.

She took her cloak from its peg by the back door. After her soaking and subsequent chill, she'd had it cut from sailcloth, soaked in oil and dried to a stiff texture that kept off wet, including the worst of the spray thrown up as the boat cut through the water. It had a basic hood and tied under her chin: inelegant, vaguely ridiculous, practical. Her gloves were an old pair reserved for collecting; she'd dispense with them later. Salt water spoiled the leather and she didn't want to have to ask Jack for money to replace them.

She checked the hunting bag; there was sufficient old newspaper from last time. She looped the strap over her head. 'I'm leaving now,' she called. 'Tell Mother I'll be back by mid-morning.'

'Wait.' Kate went off to the kitchen, reappearing a minute later. She tucked a folded napkin into Ellen's pocket. 'A bit of oatcake,' she said. 'You haven't had any breakfast.'

'Come, Lettie.' The dog trotted in, nails clipping over the stone flags.

'At least you have that creature,' Kate said.

'I'm not sure how threatening she is.'

'She's better than nothing, Miss. She can nip at an ankle, and bark loud enough. If anything happened, she'd make her way back here.'

Mr Barry waited by the shore. She climbed in easily now, Lettie under her arm.

'A fair morning, Miss. Where to?'

'Whiddy.'

They pulled away, cutting into the water. Barely any wind; the boat glided across the surface as though on ice. Once out far enough, Mr Barry steered southwest and they made good progress. She marked the landscape as it passed, mostly low-rising hills of bald rock, patched with scrubby vegetation. Here and there a settlement amongst the trees, though quiet as if deserted.

Lettie jumped from the boat first, and turned circles while she waited for Ellen. As the boat pulled out again, leaving them on the strand, Mr Barry raised an arm in acknowledgement. They'd fish while she worked, collect her before the sun peaked.

The wind licked at her hair and face. She stood for a while, listening to the music of the shore – the cheeps and pip-pips of plovers and oystercatchers, the rattling squawks of terns, the gently lapping water. A pale sun glimmered behind the mist. Debris lay scattered across the rocks, thrown up by the sea. Walking down to the water, she undid the thick leather belt around her waist – another castoff from Jack – and rebuckled it, securing her hoisted dress and petticoat. She folded her gloves in the bottom of the bag and began her way down the beach, just beyond the waterline, looking down as she went. Rolling her sleeves as far up as they would go, she pulled out her knife and plunged her hand in. By now she was used to the cold water. She no longer thought in terms of *seaweed*, or *wrack*. Rather she categorised the amorphous, undulating plants that grew under the water, exposed by tide or tossed onto the beach by storms, according to the genus *Fucus*, in all its varieties, each differing in colour, habit and function, though connected, as members of a family were.

Tugging on a clump, she slipped; she windmilled her arms but still fell with a cry and a splash. She found her footing, stood up. The bottom of her dress was soaked, her sleeves up to the elbows. She held up her skirts, stumbled onto the shingle and padded further up the rocks, dripping as she went.

Here, strewn across her path, lay a specimen she'd not seen before.

She squatted, forgetting the damp around her legs, and seized it, turning it over, focusing her eyes across its parts: midrib, fronds, curiously waved edges. The fructifications resembled those of *Fucus hypoglossum* ... Stupendous luck: the pods burst before her gaze, displacing the seeds either side of the rib. Yes, it greatly resembled *hypoglossum*, though larger, and a particular red colour, like aged claret.

She released her skirts to flap in the wind with a sound like beating wings. She laid out the specimens in a circle, placing her last discovery at the centre, and found a flat rock to sit where she rested, making notes and sketching.

Fumbling for a pencil her fingers closed instead on the wrapped oatcake.

Lettie roused herself from where she'd been napping and came over to stare.

'Are you hungry?' Ellen said. 'Of course you are. As always.' The dog's ears pricked to her voice. Her head tilted to one side as she considered her chances. Ellen threw her half. 'Don't tell Kate.' Lettie swallowed the cake in two long gulps and licked around her mouth and jaws for crumbs.

Ellen got to her feet and scrambled around the shingle, wrapping the partly dried *Fuci* in newspaper.

'Miss Hutchins, is it?'

She whipped upright. A Bantry man she knew, Michael Murphy, with a woman and three children behind him on the rocks: his family. The woman had a basket strapped to her back.

'Good morning, Mr Murphy.'

'The weather's holding,' he said, cautiously. He had a bluff, good-natured face.

'That it is. Will it brighten later, do you think? Quiet, Lettie!' The dog had stiffened, growling. At Ellen's voice she stopped.

Mr Murphy looked into the sky. 'There's a breeze might lift it.' He nodded, in the general direction of Ballylickey. 'Mrs Hutchins is well? Mr Hutchins?'

'Very well. Are these your children?'

'Yes, miss.'

'You're lucky to have so much help.'

'I am.' Mrs Murphy stayed at a distance. Her hair, loose and uncovered, the colour of pale rust, blew around her head. She might not speak English as well as her husband, if at all. The children clustered near their father, watching Ellen with naked curiosity. Their feet were bare.

'Are you collecting wrack, Miss?'

'Yes,' she said.

'They say you're interested in it.'

'I am.'

He scrunched up his eyes, as though this were a great mystery. 'And what would you be collecting wrack for, if I might be so bold?'

She remembered her bare hands and rummaged in the bag for her gloves, feeling her way under the paper. 'There are educated men, in Ireland and England, who have need of specimens.'

'Do they pay?'

'No, no.' She smiled. 'I send one sample of each, dried. With the seeds, if I can find them attached.'

Drying *Fuci* had proven another challenge, though less tedious than using the box of sand. Once home, she floated each plant in a large shallow dish of water to extract the salt, separating and expanding the delicate branches. She could then slide a stiff piece of writing paper into the water underneath, and remove it slowly so that the plant remained in its expanded state. After trial and error, she'd worked out the pressure necessary to apply the blotting paper and weights, then how to preserve the sample on paper, suspended in all its beauty and colour as though still under water.

Before she could explain some of this to Mr Murphy, he said, 'For what purpose?'

'Well. For examination, and study. To understand how a plant connects to the rest of the natural world. For example, to learn how it ...' She paused, about to say 'reproduces'. 'To learn how it thrives, or why it does not.'

'Ach.' A dismissive sound, unbelieving. He stared beyond her, at the sea. 'They just grow, Miss, as and how they will, as they've always done.'

'Certainly. But they still might teach us something about how plants generally prosper, or fail.'

'They're wild, Miss. God given. It's for Him to know the workings of his creation.'

'Of course.' She stepped forward in her eagerness to explain. 'But having created the world, would God not want us to appreciate it? Have you never wanted to know how a thing functioned? A desire to understand why it does what it does? And if you did understand, might it not assist you in your dealings?'

His face creased in polite concentration. 'If you say so, Miss.' The children continued to squint at her, open mouthed. No doubt they knew more about the tides, the weather, the landscape, than she would in a lifetime. She felt suddenly foolish. Flushing, she pulled on her gloves, forced a smile. 'You're gathering wrack for the fields?'

'If we're not disturbing you, Miss.'

'Of course not.' The children now looked out to sea; she followed their gaze. The boat had come into view. The sun was almost overhead, the morning spent. 'I must get home, Mr Murphy. I wish you good day.'

'Good day to you, Miss.'

Ellen nodded at his wife and got a tip of the head in return. Her face remained blank. Ellen walked away as briskly as she could manage across the deep shingle, towards the waiting boat.

'Stop here, please.' They were passing by a crop of islets, rocky, grassy outposts off the shore, inhabited by naught but seabirds and scrubby vegetation. 'Can you get close?' she asked.

'We can try, Miss.' By now, Mr Barry was used to these random forays off course; his expression was, as always, disapproving but indulgent. He pulled the boat in, slowing it to a bobbing drift, manoeuvering her as best he could around the rocks.

'I'll need to climb on.'

Without a word he angled the boat forward, skilfully avoiding scraping the sides, positioning Ellen where she might best get a dry foothold.

'Hold steady.' He held out his arm, solid as an oak branch. 'Take care!' She used his arm as leverage and stepped onto the rough shore.

Here she was queen, of a tiny kingdom. How peaceful it was, and disorienting. An ill-judged step might send her tripping into the water, or onto bare rock to break a limb, or her neck. She climbed about, searched clefts and fissures. Nothing.

Mr Barry's voice floated to her, borne on the stiff wind: 'All right, miss?'

'I'm coming back.'

He stepped halfway out of the boat, saw her safely on board, launched them towards home.

As they rowed, she leaned forward, shouted against the wind, 'He may have been mistaken.'

'You'll find it, Miss. Though it may not be what you're expecting.'

'Did you know him?'

'Aye.'

'A terrible thing.'

'Aye, Miss.' He looked away, as if avoiding any more discussion.

She sat back, folded her hands, shut her eyes to the wind.

Once close enough to the house, Lettie scampered away and disappeared from view. Ellen followed slowly, the weight of the bag digging into her shoulder. More activity at this hour around the yard: the sound of axe blows from the woodshed, the scratching of a broom against the cobbles. The back door lay open. She pulled off the sodden boots, pulled on her slippers.

Joanna was on her knees, scouring the kitchen floor. Her skirts were dark with wet. Kate sat at the table, darning clothes. Ellen stopped in the doorway.

'Come ahead, Miss.' Joanna gestured for her to cross the wet flags. 'I can go over it again.'

'I'm giving you more work.'

'No matter, Miss.'

Ellen stepped across the floor, filled a glass from the jug and drank. 'I saw Michael Murphy on Whiddy,' she said. 'He was with his wife and children.'

Kate put down the mending. Her eyes narrowed. 'Which Michael Murphy? Fisherman, or farmer?'

'Farmer.'

Her face cleared. 'He's a decent soul. What was he at?'

'Gathering wrack. He was curious about my samples.'

'There aren't many gentlewomen to be seen wandering about the beaches.'

She saw Joanna shoot Kate a furtive look. Setting down the glass she asked, 'What is it?'

'There's talk, Miss.'

'What talk?'

'Joanna's been telling me some of them around here think you're a spy.'

'Kate!' Joanna's mouth fell open.

Ellen hadn't expected this. 'Spying on whom?'

'Anyone up to what they shouldn't be. *Na Buachaillí Bána* – the Whiteboys, for instance. What's left of them.'

'What would I know of the Whiteboys? They disbanded years ago, surely?' Kate picked up the shirt she'd been mending. 'You're jesting, Kate. Or Joanna is mistaken.' Surprise gave way to outrage. 'Our family's been here since the 1600s.'

Kate frowned at her stitching. 'But you haven't, Miss. A blow in, from Dublin, some say.'

'I would be extremely distressed if anyone thought such a thing.'

'Ah, Miss, she's making it sound worse than it is,' Joanna burst out. She sat back on her heels. Her lean face shone with sincerity. As was her habit, she had tucked her hands, raw from being constantly soaked in hot water, into her sleeves.

'Are people suspicious of me, though?'

'Stupid talk, is all. And if anyone from this house hears it, they soon silence it. Most admire you, Miss. How you get to places. The top of Knockboy Mountain! Many men could not manage it.'

Kate said, 'They think you peculiar. But the gentry are allowed to be peculiar. It does no harm.'

'I can't believe I'm the cause of any sort of discussion,' Ellen said. 'I consider myself very dull.'

'Not so dull,' Kate said. 'Unusual enough, at least, for comment.' She looked up. 'There's no ill will toward you, Miss, just gossiping out of loose mouths. The family's too well thought of. Your father was the fairest magistrate they ever had round these parts.'

At times she struggled to understand Kate's motives. 'Come, Kate. Do you mean to frighten, or warn me?'

'Neither, Miss.'

'Then refrain from mentioning it. And for heaven's sake, never speak of such things to my mother.'

'No indeed, Miss.' Kate seemed affronted. 'Though she knows what the people around here are like. Better than anyone.'

'Well, mind what I say. Now I need to wash my hands,' she said. 'Then I'll take tea with a slice of bread and butter.'

'Joanna, pull the kettle across,' Kate said.

Ellen went upstairs. Her damp underclothes itched against her skin. While changing them, she wondered. Was she watched from the hedgerows, behind walls? Were her actions reported back to others? Well, she thought, let them.

A woman squatting in the grass or stooped over shingle, staring for minutes on end at scraps of vegetation – was it so remarkable? She was surely no more interesting than a sheep on the headland, standing for hours at a time in one spot, chewing on its own philosophy.

This image, less than flattering but strangely apt, brought a smile, which stretched into a yawn: the early start catching up on her. Yet she couldn't resist diving into Withering before resting, searching through her existing specimens to compare with the large red seaweed she'd collected at Whiddy. Now she was certain she'd never found anything like it before.

Dawson Turner would know.

Jack called her from his room. She pushed the door open.

'Will you read to me awhile?'

'Oh, Jack. I'm so tired, I doubt I could stay awake five minutes.' His face fell. 'Tomorrow, perhaps.'

'You've been out all day.'

'Yes, making the most of the weather.'

'Did you see those mountain men, the shepherds?'

'Not this time. I was out on the boat.'

'Ah.' He sat back against the pillows, a dreamy look on his face. 'When we were boys, we roamed the mountains and beaches like natives, wild as goats.'

'I wasn't allowed your freedom, being a girl.'

'No. But you have it now. You are more free than any of us.'

She laughed. 'Hardly.'

'Yes, if you think about it. Arthur is bogged down by his children and that estate. Emanuel in London, where he does not want to be, no more suited to the law than I am. Samuel with his textbooks, the weight of all our expectations upon him. I, the most unfortunate of all. And you, whom no one expects anything of, soaring above us.'

She went across to the bed, smoothed his hair back from his forehead, kissed it. 'Strange reasoning. Go to sleep.' At the door, she turned back.

'Jack? What do they say of us?'

'Who?'

'The people around here.'

'They like us well enough. Tolerate us, while they must. And as soon as they have the chance, they'll rout us out.'

'We belong here.'

'As part of an unnatural order. Here I lie in a feather bed, crippled though I am, well-fed and comfortable, while they toil and labour all their lives for nothing but semi-starvation and squalour. And you with enough time to walk about collecting plants, which they forage for bare survival.'

'Now I regret asking,' she said.

He snorted. 'We exist within a fragile system. Best to remember that.' On her way out the door, he said, 'Sleep well, sister, if you can.'

TEN

She had been wondering about Mr Turner's looks, letting her imagination play, though knowing it foolish and shallow. As if seen in a mirror that had lost most of its mercury, an image shimmered definitely in her mind. Dark hair, thick and wavy, brushed from a high forehead. Deeply set, twinkling eyes, with a permanently searching expression. His cravat askew at the end of a busy day, ink on his fingers. A coat sleeve of fine wool, soft as moss and smelling of new leaves.

Fanciful, she knew. Perhaps he was fat and squat, had red hair and bristling whiskers, puffed and wheezed when he walked. Or he was six foot and thin as a reed; or one of those small, prancing men with weak chins and eyes like boiled eggs.

She constructed an ideal, which moved and spoke, smiled and frowned. Said her name – Miss Hutchins – in a voice that was crisp, musical, tinged with the particular tones of his part of the world, though in fact she had never heard a Norfolk man speak, and did not know the accent …

Said her name: Ellen.

Did he wonder what she looked like? She could answer: tall, for a woman. She ducked her head coming in the back door at Ballylickey; Kate barely came to her shoulder. She shared her brothers' looks, though the Hutchins' features were perhaps less satisfactory in the female than in the male. She never occupied herself much with the looking glass, except to check if her hair was askew from rooting around in brambles, or if she had mud on her cheek from digging out plants. She chose her clothes to withstand the weather, rather than to flatter her figure, or to pose in as an ornament in a drawing room.

She felt somehow that Mr Turner would think as she did, that there was greater value in the modest, the hidden. The soul of a thing, shining within: quiet, withholding its mystery.

ELEVEN

Looking up at the creak of the gate, she dropped the trowel.

'Ellen. I find you hard at work.'

Tom Taylor.

She had written to invite him, more than once; he had responded with enthusiasm but had never committed to a date of arrival. Now he stood there, holding his hat, as if conjured from the smoky November air, looking at her with a bemused expression – on her knees in the mud, skirts looped up with the mannish belt, a battered straw bonnet on her head, red-faced and sweating from hacking away at dead vegetation. She got to her feet. Her petticoats showed; she had no choice but to loosen the belt and shake out her skirts. Why hadn't they asked him to wait in the house? In her imaginings of this meeting, the scene would be composed, dignified, she perhaps with a book of poetry in one hand, pleased and detached as he walked in, effusive, full of regret … Never had she planned to be caught so unawares, without a chance to ready herself, present a measured reaction.

He had surely noticed the instant gladness in her face, before she had a chance to damp it down.

'They offered to send for you,' he said, apologetically, 'but I wanted to see your garden.'

'There isn't much to see, this time of year. I'm tidying up for winter. If you return next summer, you will be better rewarded. If I manage to make anything grow, in spite of the efforts of the rabbits and slugs. They seem determined to thwart me.' She was babbling.

'I promise to return when it's a veritable Eden.' He stepped forward.

'Wait there, please. It's exceedingly muddy. See, I have my brother's old boots on. I shouldn't like you to spoil yours.' She came down the garden. 'Excuse my untidy appearance.'

'The fault is entirely mine, arriving without warning,' he said. He'd filled out, his whiskers grown longer. 'I thought to continue straight on to Kerry, but decided at the last moment to call on you.'

'You're welcome, of course. We have so few visitors I've got used to dressing for comfort rather than society. Have you come from Dublin?'

'I stopped in Cork for a few days. I hadn't fully appreciated how far from civilisation you are. Though a beautiful situation.'

'The country is bleak this time of year,' she said. 'Did you meet my mother and brother?'

'Briefly.'

'Well. Now that you're here, I have a hundred questions you can help me with.'

'Straight to business, as always.' He smiled. 'Thank you again for the samples you've sent. You've been exceedingly generous.'

'Happily, I've been able to collect sufficient quantities to share amongst my friends. There's a treasure of *Fuci*, particularly, in the coves around here.' As they came to the front door, Leonora appeared.

'There you are,' she said. Her eyes met Ellen's. *I tried to delay him.* Ellen nodded: *it's all right.* Leonora held out her hand. 'Won't you come inside, Mr Taylor?'

'I'll take a turn around the lands first, if I may,' he said.

'Unfortunately Jack can't accompany you, much as he might like to.'

'I wouldn't trouble him,' he said, already walking off, waving his cane. 'I can find my own way. I have a desire to walk down to the sea.'

Leonora followed Ellen into the hallway. 'We couldn't stop him going to find you himself. What resolve that young man has!'

'He must take us as he finds us,' Ellen said. She pulled off the detestable bonnet.

'Your hair, your dress!'

'Mr Taylor pays little heed to my looks.' Though, she thought suddenly, he might have walked away precisely to give her an opportunity to tidy herself.

Upstairs in her room, she shook out her hair. It resisted the comb; she tugged at it, damped it down with water, pinned it to something like order. Another woman, she knew, would cherish such hair, wash it, brush it to shining smoothness twice a day. She made a face in the glass. Too late to do any more with it now. About to leave the room, on impulse she went to the cupboard where she hung her few dresses. She lifted down the most decent amongst them – a blue striped cotton preserved for indoor wear – and held it before the mirror.

The skin around her eyes looked puckered, her cheeks wind-chapped. She shook her head at the woman in the mirror, hung the dress back in the cupboard. How could he view her as anything but a likeable, fusty oddity, due no more than common courtesy based on old acquaintance and distant kinship, when she looked like this?

It would take more, she knew, than a change of clothes.

He lay back in his chair, drinking tea and eating fruitcake. One of his legs was crossed over the other, and bounced slightly. Leonora, nested in shawls, alert and composed as a little owl, was regaling him with stories of the Protestant families of West Cork. Social hierarchies, lands owned and inherited, worthy landlords who invested their energies and wealth in the betterment of their tenants, and those who left their estates to rot while they frittered away the family fortune in Dublin or London – Leonora had intimate knowledge of them all.

'Of course,' she said. 'Things are not what they were in my husband's time. Estates auctioned to pay debts, woodlands sold off, land divided amongst sons.' She dabbed a cake crumb from her lips. Her little gloved hands curved inwards. They had been causing her pain lately. 'The old families usurped by new money.'

'Some are glad to shed the burden of these great estates,' Tom said.

She sat up, instantly agitated. 'That is beyond my comprehension. How could land ever be a burden? The Hutchinses, on both my side and my husband's, held land here since not long after Cromwell's time.'

'Ultimately, whether a family thrives or disappears depends on the vagaries of one generation to the next.'

She shrank back into her chair. 'True, sadly.'

'A family may distinguish itself in ways beyond ownership of land,' he said. 'A single member might elevate it.'

'By marriage, of course.'

'I'm thinking of other achievements.' He looked at Ellen. 'I'm keen to see how your herbarium progresses, Miss Hutchins.'

Leonora fluttered her hand at them. 'I know you young people are keen to talk about plants. Ellen is in sore need of a knowledgeable companion, Mr Taylor.'

'We won't be long, Mother,' Ellen said quickly.

On the way up the stairs, he said, 'It's a fine old house.'

'I've grown to love it. Of course, it's not so elegant as 16 Harcourt Street.'

'No, it's a different prospect entirely. This house makes a virtue of function: to provide refuge from wind, rain and isolation; modest, in recognition that it could not and should not dominate its setting. It surely reflects your family's good taste.'

On the upstairs landing, Ellen opened the doors leading on to the balcony. The distant mountains stood purple, a
backdrop to the soft gleam of the bay. A passing boat slit the
water, casting rippling shadows in its wake.

'Harcourt Street does not have such a view,' Tom said. 'I
wouldn't trade this for the finest dwelling in Dublin. What
peace it would bring the soul, to wake to that each morning.'
They stood in silence for some minutes. At the same moment
they turned and went inside. Ellen led him down the corridor
and opened the door of her study.

Standing at her desk they went through the loose sheets
of the herbarium, each page neatly marked with name, date
and place of origin. Tom made useful, thoughtful comments,
and responded to Ellen's questions, 'Though, you are on your
way to becoming a superior botanist to me.'

'I have no need of flattery.'

'And having some idea of your nature, I wouldn't offer it.
All the same, I state the truth. Stokes and Mackay, diplomatic
as they are, agree.'

Flustered, she paused in her turning of the pages. 'It's
kind of you to say it.'

'Kindness has nothing to do with it. It's plain fact.' He
carefully laid down the sheet and lifted the next. The red *Fucus*.

'What is this?'

'That particular plant is the first identified of its kind,
or so I believe,' she said. 'Mr Turner of Yarmouth thinks it
resembles *Fucus membranaceus* though in my view it is more
like *Fucus hypoglossum* … he wants to include it in his revised
History of Fuci.'

'Will Mr Turner acknowledge you as the finder?'

'I asked that he not use my name.'

'Why not?'

She looked down. 'It wouldn't be decorous.'

'Who says so?'

'Is it not the perceived wisdom?'

His eyebrows furrowed inward. 'Who will Mr Turner say found the plant?'

'A lady from Bantry.'

'Cryptic, and distinctly lacking, in my opinion. I think it a shame that you should not receive credit for your findings. Surely Mackay would agree?'

It was Mr Mackay she had charged to write to Dawson Turner and say she didn't want her name published. 'He respected my feelings on the matter.'

'I think you should reconsider.'

A brief silence. 'I would have to ask my brother, Emanuel.'

'Surely he would consent. It's a matter of science, after all. What could be the objection?'

'There are things a woman has to consider, that a man does not,' she said.

'With regard to botany, what exactly?'

'A woman, once judged to have acted immodestly or improperly, has no way back. Her reputation is fragile. Her future …' She stopped. 'Her future hopes might be impaired.' A pause. Distracted, she had prised away a blob of hardened wax from the table; now she crumbled it in her fingers.

Tom stared at her. Then he shrugged. 'As you say, your brother is the best person to seek advice from on the matter. I still believe you should receive due acknowledgement for your work.' He began again to leaf through the herbarium. 'Here's a beauty!' Ellen, glad of the change in subject, followed his gaze.

'*Fucus bifidus*,' she said. 'A rare type.'

He moved beyond the *Fuci* and came to the flowering plants. 'The quantity you have collected is astounding,' he said.

'The benefits of rising early and having little other diversion.'

'What's this?' He peered closer. '*Erodium moschatum*.'

'Musk stork's-bill, rather rare in these parts … I found it around the walls of Whiddy Castle. I did find it once before, on Howth Head, to the delight of dear Dr Stokes. I still have the specimen, pressed in a book. For some reason I kept it all this time.'

'You realised you had found something exceptional. It's what sends us into marshes, down dark valleys, over impossible rocks and bogs ...' He shook his head. 'We are permitted an occasional inkling of sentiment, however at odds with the practical rigours of science.'

'I agree, within reason.'

'Reason. Always reason, dear Ellen.'

'What else is there?'

'Well,' he said. 'I have an impulse to see more of the locality, before I travel on. Would your mother be amenable to putting put me up for a few nights, do you think?'

'I'm sure she'd be delighted. You seem to have charmed her completely.'

'I trust I find favour with all the inhabitants of Ballylickey House.'

'Why would you not?' she said. It came out more boldly than she intended.

'I'm forthright, as you may remember.'

'That's only to your credit.'

'You're not offended by what I said earlier?'

'I appreciate your frankness.' She busied herself reordering the herbarium, filing the separate sheets in the drawers of the bureau. Then she paused, looked at him. 'I will consider what you said. At the least, I'll write to Emanuel for his opinion.'

'Let's hope he sees the matter as I do.'

She sighed. 'I wish the rules were clearer.'

'Rules?'

'Those by which we must live. Spoken and unspoken. People hint, and offer opinion, but it's all based on their own notions.'

'I cannot see how you can do wrong by trusting your own feeling.'

This fluctuated from one day to the next.

'If I were a physician, I should be surer of myself.'

'A pity that cannot be.' A short silence, then, 'Perhaps you can act as my guide for a morning. If you can be persuaded to leave off toiling in the mud for one day.' He smiled.

Jackdaws, startled by the sound of human voices, flapped from the high gables and chimneystacks of the old manor house.

Tom indicated the ivy-choked walls, stark against the sky. 'It appears Tudor, the oldest parts at least.' The building had no roof or window frames. Weeds choked the spaces between the remaining flagstones.

'Reendesert, the locals call it,' she said. 'Once owned by the O'Sullivans. A ruin since Cromwell's time, apparently.'

'It was well fortified once. You can see the bartizans and gunloops. Evidence of a violent past. Can we walk around it?'

She hesitated. 'I believe the house is now owned by the Bantry estate.'

'Lord Bantry won't begrudge us. I'm sure O'Sullivan himself would provide an Irish welcome were he here. Isn't there an Irish custom of allowing visitors to freely tour your land?'

'That can be misjudged, resulting in oaths and gun blasts.'

'Not amongst gentlemen. Come. We've not been shot yet, nor chased off by dogs.'

Around the ruin was a confusion of fallen masonry and foliage, yet here and there the remnants of stone pathways showed evidence of a garden of sorts, random bones of something once formal, organised.

'There's an atmosphere, as if the place senses it has been abandoned,' Tom said.

'Now it is you who sounds unnerved ... Just another ruin, Tom, in a land of ruins.' They wandered on and found a corner formed of two remaining walls, separate to the main house. The quiet was deeper here. Even the wind had muted.

'An outbuilding of some sort?'

'Or what's left of the walled garden.'

'Perhaps. Look at the variety of moss.' Tom walked slowly back and forth, stopping to point, examine, exclaim. 'Here, Ellen. *Tortula rigida. Tortula cuneifolia.*' On the other side, lower to the ground: '*Tortula muralis.*'

She joined him. Their heads bent close together, their fingers brushed against each other over the soft cushions of moss. His face had the animated, focused intensity she remembered; this, she realised, was when she liked him best, when she felt they were in harmony – equals, bound by a curiosity that to outsiders would surely appear like a kind of madness.

She laughed softly, he looked up. 'Something amuses you?'

'I'm remembering when we walked in the Phoenix Park. You became greatly excited then, by a moss.'

He smiled, leaned back against the wall, arms folded. '*Tortula papillosa.*'

'You remember! I could not, for it had little meaning for me then.'

'That walk was memorable, for many reasons.' She said nothing. 'I took liberties with your person, you were furious with me. For a moment, at least.'

'My foot!' He laughed. 'I couldn't decide if you were supremely confident,' she said, 'or simply without manners.'

'Both,' he said. 'I thought you very tolerant, and self-assured.'

'I, self-assured?' She shook her head. 'I was quite in awe of you, I think.'

'Was?'

She shook her head. 'I was a young girl. Come, enough reminiscing. Let's hear what you think of these mosses.'

He smiled, took the field microscope she offered. As always, he had sure, instant opinions. She generally agreed but dared to question him once or twice, forgetting her reticence as he held to his argument, until she became as heated as he was. 'We can settle it by putting it under the better micro-scope,' she said.

He smiled, satisfied. 'Was it not worthwhile to wander into Lord Bantry's domain?'

'I admit it was.'

From the road, the sea was visible through a tangle of vegetation. Tom strode about, thrashing his cane at the bri-ars. He shouted, 'Here. I believe I see a way down to the beach.' A narrow, overgrown track wound downwards; soon they found themselves on shingle that stretched to where great, grey waves crashed in. They walked as close to the edge of the water as they dared. Ellen bent to root amongst the stones. 'Remember I wrote to you of the fisherman, O'Sullivan?'

'The one that drowned.'

'Yes. The tree-type plant …'

'You've found nothing like what he spoke of?'

'No. Unless it's a new species …'

'You'll keep looking?'

'He saw it. It exists.'

'Still, it's the word of one uneducated man.'

'He made an impression on me, that I can't forget.' She stood, wiped her hands on her dress.

'You will stay here,' he asked, suddenly, 'so long as your mother lives?'

She looked up, surprised at the change of subject. 'There's no one else to take care of her.'

'You could take her with you. If your circumstances changed.'

'She would not go anywhere else. Arthur asked her to live with them at Ardnagashel. She refused.'

Ulva ramulosa. She threw it down, she had plenty.

'Elderly people grow attached to what they know.'

'Not just the elderly. I believe Jack would find it impos-sible to leave. Though he says differently.'

'Understandable. He's known little else.' His voice had lifted, snatched away in the breeze; he gazed out to sea. 'Poor soul. To end up like that from a schoolboy accident.'

'Of all of us, he has been dealt the cruellest hand.'

'But you would consider leaving, if you could?'

'I could be happy elsewhere, yes.' The first drops of rain studded the stones. She tightened her shawl about her shoulders. 'And you?' she asked, lightly. 'Will you settle in Kerry, eventually, as you've said?'

'Kerry. Or somewhere on this coast.' He waved towards the headland, the ruined manor house. 'A setting as fine as this, for instance.' He shrugged, scuffed his boot against a rock. 'Or I may decide to stay in Dublin. I am pulled, like the tide, forwards and back.'

'Unlike the tide, you have free will.'

'You told me once, we are not all free to decide our future. Even those who appear to live without constraints may have them.'

'My circumstances then were far different to yours. From my viewpoint, you have every advantage in life.'

He nodded. 'With good fortune, though, comes responsibility. My profession, for example – that was my father's wish. Now I must prove equal to it.'

'You feel that a burden?'

'In some ways. At any rate, life is at times a dizzying prospect.' She said nothing to this. He put out his hand, peered upwards. 'Usually some outer force compels us to act. Such as the rain, which seems determined to drive us home.'

She stumbled over the rocks; he took her hand.

No reply came from Emanuel to her letter, nor to those she sent after. Eventually she wrote to Samuel, asking him to remind his older brother that she needed a response to her question. Had he stuffed her correspondence, unopened, unread, deep into his coat pockets – as was his wont – to moulder there, forgotten, for months? If he had, her words, so carefully chosen, would be as wasted as if she'd shouted them into the London wind.

A month later Dawson Turner wrote again asking if he could use her name as the discoverer of the red *Fucus*. That night she woke in the early hours and lay wide-eyed and fretting until the servants began stirring at six.

'What ails you, Ellen?' Leonora asked.

Evening, after dinner: she had been trying to read. The words shimmered to a jumble and slid away, so that her gaze continuously wandered upward and rested on a bare patch of wall. At her mother's voice, thin and high as if coming from a distance, she blinked.

'I'm tired, that's all.' She shivered. 'And I feel cold, as though I sat permanently in a draught.'

'Have you received unpleasant news?' Leonora's tone was mild.

'No,' Ellen said. She lifted her book again, eyes roving the page. Where had she left off? She seemed to be reading the same sentence over and over. Jack lounged on the sofa, his face masked by a six-month old copy of *The Freeman's Journal*.

'But there is something,' Leonora said. Ellen gave up, dropped the book, pressed her hands against her temples. 'Have you headache?'

'Some.'

Yet she didn't move. Jack scratched himself, flipped over a page. Leonora took up her crochet. The needle resumed clicking, dancing the thread in loops.

'What would you say to my name being published? As the discoverer of a plant?' Ellen's voice, breaking the silence, startled even herself. Jack lowered his newspaper. 'In a botanical journal. Mr Turner, my correspondent from Yarmouth, is writing the article. I wrote to ask Emanuel if he approved, but he doesn't reply. I must give Mr Turner permission or say no once and for all.'

How tangled her concerns sounded, magnified, yet trivial. An unopened letter shoved in a coat pocket. Damn Emanuel.

Why should he have such sway over her affairs, when he clearly cared not a whit?

'Might you become well known?' Leonora asked.

'Only by those in botanical circles.'

'By putting your name to this discovery you'll be forever known as the lady botanist, Ellen Hutchins, of Bantry Bay, Ireland?'

'Yes.'

'Did you consult with Thomas Taylor on the subject?'

'His opinion was definite. He says I should agree to it.' He'd written since, news of his studies, his own collecting. After signing off, he added a postscript: *Let not your reticence do an injustice to your talent.*

'What's your own feeling on the matter?'

Exasperated: with her own indecision, as much as with Emanuel. He could have resolved the matter, either way, made it easier by taking responsibility. Lying awake in the dark, staring at the ceiling, she'd admitted the truth to herself. She wanted her name credited. She wanted to say 'yes' to Dawson Turner, instead of affecting a modesty she didn't feel.

'I shouldn't like to do anything that my brothers might disapprove of,' she said carefully.

'Jack?' Leonora said.

'Yes?' His voice was sleepy.

'What say you to your sister's dilemma?'

'What dilemma?'

'Have you not been listening?'

'I don't have Emanuel's permission,' Ellen said, 'Though I wrote ...'

'He didn't reply. I'm not surprised. You must know by now he only has regard for himself.'

Leonora shook her head. 'Unkind, Jack. Unkind, and unfair.'

'Is it?' Ellen asked. 'Then why not answer my letters?'

'He meant to, surely, and forgot.'

'You make excuses for him, as always.'

'I give him the benefit of the doubt, as I would give you all. He has many cares.'

'What cares?' Jack said. 'Betting on horses, starting arguments in taverns over an independent Ireland?'

'Jack,' Ellen said. 'You go too far.'

Leaning on his elbow, he pushed himself to sitting. His hair stuck upward; his rumpled shirt had come undone at the throat. 'As the master of the house, I give my permission. Let Ellen receive credit for her hard work. Why not?' He reached for his wineglass, raised it towards her. 'After all, she's the pride of Ballylickey.'

Was he being sincere, or sarcastic – or both? She exchanged a look with her mother. *Don't argue with him.* Leonora nodded: 'And you say Mr Turner is a man of good reputation?'

'Mr Mackay speaks highly of him, and Mr Mackay is a friend of Dr Stokes.'

'That settles the matter,' Jack said. 'Really, Ellen, you make a fuss over very little.'

'I confess to having been in a state of turmoil.'

'Well, no more.'

'With that in mind, I must go to bed. Mother, will you come too?'

In Leonora's room, the fire had almost gone out. In the trembling light, Ellen unpinned her mother's cap, rolled down the thin rope of hair. She took up the hairbrush, enjoying as always its comforting heft. Solid silver: the handle worn to a shine over decades.

Leonora said, 'You've brightened this house ever since you've returned. With all your endeavours.'

Ellen couldn't see her mother's face. She slowly, carefully stroked through the fine grey strands. 'Would Jack agree?'

'He more than anyone. Even if he might not ever say it.'

Later that week, clambering over the shoreline near Glengarriff Harbour, she found more of the red *Fucus*, glowing against the black, wet rock. She would offer it to Dawson Turner through the post, for his book: give her name – gladly – to its discovery.

TWELVE

The next fine Saturday they went into Bantry. Leonora insisted on first visiting Ellen's father's grave. They stood in silence for a time, as the wind whipped and moaned amidst the old stone markers and lifted their skirts. Leonora wiped her eyes.

'Let the men wait here while we do our errands,' she said.

Dan and the stable lad Patrick stood outside the wall with the horses. The road from Ballylickey lay impassable to carriages, entire parts flooded by the recent rain, the rest bogged in a quagmire. The only way of reaching Bantry was on horseback, riding pillion behind the men. Leonora justified it, saying, 'It's that or never leave the house.'

They set off down Chapel Hill, Ellen guiding her mother across the worst of the rough paving. Slowly, then, past the post-office; the inn, the Bantry Arms (a mean-looking place with little sign of life at this time of the year except for a thread of smoke coiling from the chimney); Market House, one of the few handsome buildings in the town and the focus of much of its commerce; the flour factory; the brewery, Bantry Mills. They came eventually to the quays, lined on one side with solid, neat houses. Leonora knew some of the inhabitants but would not call unannounced.

The harbour, at least, was busy. Boats pulling in, boats setting out: mainly hookers that fished the waters of the bay and beyond for hake, mackerel, herring and sprats, from Monday dawn to Saturday nightfall. Ellen and Leonora picked along the quays, watching their feet, avoiding piles of scales and fish guts, discussing each boat, its name, its owner, what they knew about the man and his family, the town and its history.

'They say the pilchards have all but disappeared,' Leonora said.

'Were they as numerous as people say?'

'They spilled onto the quays like a great shining river. I expect that's why they eventually diminished, taken in such numbers.'

'Perhaps they went elsewhere, compelled by some change in the water, indiscernible to the fishermen,' Ellen said.

'A mystery.' Leonora sighed. 'All we know is, the great stream of silver came to an end.'

'But the town adapted.'

'To survive, yes.'

On the opposite side of the harbour, Bantry Estate sloped towards the shore, magnificent in comparison to the modest town that nestled at its boundary.

'I should like to see inside the house,' Ellen said.

'The Whites may yet invite us. When last we met, they were more cordial than they had been for years.'

'Why should they not be?'

A pause, before Leonora said, 'The Wolfe Tone business.'

'The French invasion?'

'The attempted French invasion,' Leonora said. She sniffed. 'Richard White, Lord Bantry as he now is, took the English side and was made a Baron for it. For a time they shunned us.'

'What had you to do with the affair?'

'Nothing.' Leonora stared across the water at the estate walls. 'Emanuel may have been very slightly acquainted with Wolfe Tone at university. I suppose they wondered if ... utterly unfounded, but ...'

'Suspicion lingered,' Ellen said. 'As to the reason Wolfe Tone chose Bantry as landing place.'

'Emanuel had nothing to do with it,' Leonora said. After a pause, 'He swore.'

'Weren't they close friends?'

'Who told you that?'

Better to leave Tom out of it. 'That was my impression,' she said, vaguely.

Leonora's face soured. 'Need I remind you that Wolfe Tone ...' She paused, lowered her voice. ' ... Need I remind you that Wolfe Tone was charged with treason, and died a criminal? Nothing was proven against your brother, nor was any accusation ever directly made against him. I take him at his word. Whatever Richard White thought.'

In the window of a house on the Strand they stopped to read a note:

> *Miss Anne Quirt, dressmaker. Alterations and repairs. Clothing made to order, for ladies and gentlemen. Enquire within.*

'She could help with some of the lighter needlework,' Leonora said.

'Kate and I manage.'

'We're not lacking in money?'

'Jack is a good manager. But we have to be careful.'

'Samuel's schooling is expensive. Which, of course, must be the priority.'

'And his shirts, suits, materials, pin money ... we make do with what's left over.'

'You have everything you need? For your botany?' As if it had only occurred to her that Ellen's work might cost money. 'You can't expect Dr Stokes to supply you with everything.'

'Nor do I,' Ellen said, stung.

'Yes, but if you need anything you must ask me. Money will be found. I can always sell something. I have jewellery ...'

'I would not allow you to do that, Mother.'

'If you needed books, for example.'

'There are certainly books I should like, if money were not an object.'

'How much, for one of your friend's botany books?'

'They are expensive, and if they have plates, even more so.'

'Yes, yes, but how much?'

'Well.' Ellen gave in. 'Five pounds, perhaps.'

Leonora stopped walking. 'For a book?'

'Yes.'

'I'm not sure my poor rings would fetch as much.'

'Please, no more. I could never ask you to sell your keep-sakes, even if they bought a rich man's library.'

Taking the same route back towards Kingston's grocer on Main Street, they passed close to the other main thoroughfare that led out of town. The houses here were mere cabin-cottages, dismal dwellings that seemed to have erupted from the earth like wens on skin, in the process of slumping back into the mire. Life a-plenty though: children running between houses, women talking in clusters, their heads covered in shawls. A yellow haze hung over the street, and the dark, peaty smell of turf smoke and bracken. The occupants worked as fishermen, or in the curing sheds. Once, Leonora said, in Ellen's father's time, the Hutchins family had owned some of these 'fish factories'. No more: many of them were now derelict.

At the post office on Main Street Ellen said, 'I must enquire if there's anything for us.'

They stepped inside. The weak-chinned, stubby post-master, Mr Clerke, greeted them and began to search the bank of cubbyholes. He nodded – 'There is indeed a letter, Miss Hutchins,' – with exaggerated civility, and peered at the stamp. 'Another from Yarmouth. One shilling ten pence, please.' She had the coins ready in her glove and he handed the letter across. She recognised the now familiar sloping handwriting; placing it carefully in her pocket, she said, 'Come, mother, we have other errands. Thread, candles, salt.' Though what they had really come to town for was what Leonora called 'a breath of fresh air' – a change of landscape, different faces, words, however brief, exchanged with someone other than the servants.

Back on the street Leonora said, 'You receive a quantity of correspondence these days. From the English botanist?'

'Mr Turner, yes. Also, Mr Mackay writes from time to time. And others.'

'I wonder, though.' The pressure of her mother's hand on Ellen's arm increased. 'It doesn't take a toll on your health?'

A recent spell of headaches, nausea, weakness in her limbs: her childhood complaint. 'I know I've been unwell of late, but it has nothing to do with botany.'

'Yet it seems to be brought on by excitement, or overexertion.'

'My work is a comfort to me.'

'That you should call it work, says everything.'

'Running the house, helping Jack, is that not also work?'

'Well. That is for family.'

They crossed the small bridge, over rushing water. A heron perched on the bank, one leg folded beneath him. A seagull circled overhead, alert, vindictive.

'Ellen,' her mother said.

'Yes?' Ellen hoped she hadn't taken her silence as a rebuke.

'If you wanted to ... leave, you know I would give my blessing.'

Ellen said, confused, 'Leave? To go where?'

'To marry, for instance. If you found a suitable match.'

A young man and an older lady came across the bridge, tipping their heads in acknowledgement. Ellen waited for them to pass. 'Is it appropriate to discuss this here?'

As if she hadn't spoken, Leonora went on, 'If you receive an offer, you should consider yourself free to accept. If your brother approves.'

The heron stretched out its folded leg with mechanical delicacy, as though stroking the air, and tucked it away again. Ellen said, 'I have never had an "offer", as you call it.'

'Have you hopes from anyone in particular?'

'No, Mother.'

Leonora frowned. 'You're handsome, you're young, still. These men come to visit you, write to you. It's obvious that you're admired. Is it such an unreasonable assumption?'

'Mr Turner is married, with a family.'

'They're not all married.'

'We're bound by our mutual interests. That's all. Of course, they're amiable, kind. Like Mr Mackay ...'

'And what of Thomas Taylor?'

'What of him?'

'Is there something between you?'

'Apart from friendship and botany – no.'

'As yet.'

'There's no question ...' She stuttered, paused. 'I'm sure the thought has never occurred to him.'

'No?' Leonora pressed her lips together. Her eyes watered against the breeze.

Ellen hesitated. 'You think ... you think differently?'

Her mother placed her hand on her arm. 'Why not? Affection between cousins, it's entirely natural. Your father and I had an affinity from the start. It makes perfect sense. You have so much in common. He's more sociable than you, but that's as it should be. He would bring you out of yourself, and your soberness would temper his exuberance ...' Her lips trembled with fervour; she took out a little crocheted handkerchief and dabbed at them, wiped her eyes.

'Perhaps you see what you wish to see.'

'He will speak, I'm sure of it.'

This was preposterous. Yet Ellen felt a sudden queer sensation, a warmth, a thawing in her breast. 'And how could I leave you, and Ballylickey?'

'That is what I'm saying, you foolish girl. You must.'

'Emanuel asked me to come back here, to be your companion. That is my duty, above anything else.'

'He's buried you here.'

'Dear Mother.' Ellen took her hand. 'Never think that. After all the years of separation, I'm glad to be near you. And Jack.' She lightened her voice: 'Even when he makes himself objectionable, like this morning, knocking the jug from the table. It's not Kate's fault if a tenant fails to pay his rent.'

'Poor Jack,' Leonora said. 'Could I blame you if you ran away with the first suitor to call at the door?'

Ellen took her arm. 'Come, we still have purchases to make in Kingston's.'

As they walked, Leonora's mood brightened. She began to chatter – better to buy more of the cheap, foul-smelling tallow candles, or less of the better-quality wax, to be eked out for longer? Ellen listened, murmuring agreement when appropriate.

Her mother had the unnerving gift of the blind person, who stumbles and stalls, trembling all the while, yet finds their way unerringly through the dark.

YARMOUTH, ENGLAND
JANUARY 1808

THIRTEEN

Sealing the package, Dawson Turner rang for the office boy.

'Go to the White Horse on Fetter Lane, ask for Captain Mangin, and give him this. Then straight back here.'

> *Mangin,*
>
> *I will avail of your offer and ask you to deliver this packet to the bookshop of Anthony Edwards, Castle Street, Cork. I trust it will cause you little or no inconvenience. Miss Hutchins will collect it herself or have someone do so on her behalf, so you need concern yourself no more with it once it reaches that office. It contains delicate plant specimens; though securely packaged, I ask that you handle it with care to prevent these from being damaged, so far as you are able. The recipient, a lady botanist of that locality, will be grateful.*
>
> *I am much obliged,*
> *Dawson Turner*

Now whenever Dawson met, or heard of, anyone whose business took them to Ireland, or government agents of any stripe, he mentally added them to his list of potential couriers. Specimens stood at least a chance of surviving intact in a gentleman's trunk, rather than being tossed around with the mail in a ship's hold. He also wanted to save Miss Hutchins the cost of postage where he could. If the arrangement with Mangin worked out, he might be able to rely on it again.

The following week he wrote to her again, hoping she'd received the packet of *Fuci* and *Sertulariae*, as well as a copy of his own *Muscologia Hibernican*. On the other side of the desk sat young William Hooker, working on his *Jungermannia* monograph, the scratching of his quill as fluid as Dawson's own. Hooker lived so close by that he could visit almost daily, escaping his bachelor rooms for the earthly comforts of the apartments above the bank, where, as he regularly proclaimed, he greatly esteemed the company, jolly suppers and congenial atmosphere conjured by gracious, tolerant Mrs Turner. 'What are you working on, Turner?' he asked, without looking up.

'Writing to Miss Hutchins of Bantry, Ireland,' Dawson said. 'You've heard me speak of her.'

'The source of so many of your splendid *Fuci*.'

'Indeed.'

Hooker put down his pen. 'Would she be amenable, do you think, to sending me any liverwort specimens that might be in her locality?'

'I can certainly ask. She's most accommodating.'

Hooker scribbled on a scrap of paper. 'These, to start with, if you will. And give her my compliments.'

Dawson added Hooker's request to his letter: *Jungermannia julacea, Jungermannia quinquedentata, Jungermannia curvifolia.*

'If they grow on her peninsula, she'll find them,' he said. 'I never knew such a plant hunter.'

The softest of knocks: 'Yes?' The door creaked open and Maria, Dawson's oldest child, slipped into the room.

'Papa, Mother says to tell you dinner is ready if you are of a mind to eat any.'

'It's Mr Hooker you must persuade. He's most industrious. Come look at his drawings.' Maria came across and followed her father's indicating finger: a page of illustrations, functionally, if not expertly, drawn. 'Are they not impressive?' Dawson said, giving her shoulder a soft pinch. She bent her head, drawing her thick hair, so like her mother's, behind her ear. Eleven years old, and quite used to conversing with her father's friends.

'They are very fascinating, Mr Hooker, and pretty, though I cannot readily tell what they may be.' Dawson smiled to himself. Hooker's botanical illustrations were peculiar – spindly, amorphous, alien. 'But will you come to dinner?' she went on in her high, pleading voice. 'Papa will come if you do.'

Hooker glanced at Dawson, his wry mouth curving in a smile. 'You're allowed sit up to dinner, Miss Turner?'

'Yes,' she said. 'On occasion.'

'She's mature enough to learn from the conversation of her elders,' Dawson said. 'As I was at her age.'

Hooker stuck his quill in the inkpot. 'I wouldn't dare thwart you, Miss Turner.'

Maria took Dawson's hand and tugged, as if to pull him out of his chair. 'See, she will not leave us, Hooker, in case we get distracted again,' Dawson said.

'You mean to come, Papa. But then you pick up a book or a plant specimen, and we must all wait, while the soup goes cold.'

He patted her hand then let it go. 'Very well. One moment while I finish this. I want to send it in the morning.' He signed the letter, folded the page, sealed it and scrawled the address: *Miss Ellen Hutchins, Ballylickey, Near Bantry, Cork, Ireland.*

Last to leave the room, he shut the door; otherwise the scant heat of the fire would dissipate into the chill hall. Early January and the Yare had frozen in parts. Thick sea mist shrouded the quays. From outside, the windows of Bank House must appear as a glowing refuge in the gloom.

Dawson's past descriptions of Miss Hutchins's situation had piqued Mary's imagination. 'What does she do for amusement?' she asked, sitting before her drawing board, filling out a sketch of their friend Charles Burney. Her stomach swelled high and compact. She had confessed to him her hopes of the child being a boy; if her joy brought with it any after-pang of anxiety, she kept it to herself. She appeared strong, infused with vigour, capable of anything. Though, as he knew, this burst of activity wouldn't last. During her six previous pregnancies, she had waned in the last months.

'Amusement? I don't think Miss Hutchins knows the word,' he said. He stood by the window, allowing himself a rare moment of idleness. Looking out, he saw that a frigate had just entered the port. 'The lady has a thirst for learning that I've rarely encountered. One has the impression that it's life and death. How she and Hooker would get along!'

'Why not invite her here?' Mary said.

'Are you well enough for a visitor?' He referred to more than the pregnancy.

'Well enough to entertain every botanist from here to Land's End. I think I can manage to provide for one modest Irish lady, who will no doubt, spend most of the time out-doors.' She paused at her drawing, considering, her head to one side.

'It might be to her benefit,' Dawson mused. 'A refined person, stuck in a backwater.'

'And I would enjoy some female society for a change.' Her pencil rasped across the paper. 'I wonder why she hasn't mar-ried, if she's so personable?'

'Lack of opportunity, I presume. The family leans on her. They have a firm grip on her as the only daughter. I have a sense …' He paused.

'Yes?'

He shook his head. 'Nothing. Supposition, that's all.' He looked again out the window. The arrival of the ship had brought forth a swarm of men to action: running, shouting.

After a moment, she said, 'Share, husband, please.'

'Well. I have a sense of some deeper malaise, perhaps.'

'In Miss Hutchins?'

'Her isolation is profound. There are surely few ladies of her equal in the vicinity.'

'She seems to have aroused your protective streak,' she said, good-humouredly. 'Is this the beginning of another one of your platonic infatuations?'

'There's an intensity to her writing that's almost palpable. A tension that's bound to break.'

'Ask her to come,' Mary said. 'Now, when the year still lies before us.' She held up the drawing board. 'Look. What do you think?'

'Burney, to the life.' And it was; his portly, kindly countenance in profile, his essence captured on paper. How talented Mary was. Drawing had brought her back from grief. That, and the child within her belly.

BALLYLICKEY, CO. CORK
APRIL 1808

FOURTEEN

Like a spider she wove across the peninsula, laying a marker in each place: a plant, named and recorded.

Reeling in the wind on the summit of Knockboy for *Spergula saginoides*. Wading into the river at Ballylickey to pluck *Chara vulgaris*. Plodding through a bog ditch, cold water leaking over the top of her boots to claim *Utricularia minor*.

On Whiddy Island *Parietaria officinalis* grew on the castle walls; she dug its roots from the crumbling stone. On the seashore below Blue Hill, the red fruits of *Solanum dulcamara* glistened amongst its late flowering blossoms; she scrubbed her hands well afterwards for these berries were poisonous. Scrambling through a hedge at Gurteenroe, briars and twigs snatched at her hair and scratched her arms: no matter when her fingers closed on *Arenaria trinervia*.

On Hungry Hill – not a hill at all, but a mountain, and a formidable climb even without the nuisance of skirts – she found *Rosa spinosissima*. The quiet at the summit of Priest's Leap would remind even the strong minded of the foolish story that gave the mountain pass its name – a priest on horseback, fleeing a party of soldiers, survives a miraculous leap so desperate it imprints the horse's hooves forever in the rock. Legends and superstition were rampant, but she cast them from her mind, ignoring the whining of the wind, and gathered *Saxifraga umbrosa*.

She was well aware of how eccentric she looked, squatting in a field at Laharn to gather *Mentha sylvestris*, or roaming Eagle Point for no better purpose than to secure a good specimen of *Anthyllis vulneraria* (Common kidney vetch; she now knew that *vulneraria* meant 'wound healer'). Dr Stokes had sent her a Latin dictionary and she learned new words daily, though so far restricted to botanical terms. The folk she met either stared brazenly, almost tripping over their feet with curiosity, or looked pointedly ahead, as if bent on avoiding her eye. They were not used to seeing a lady, bonnet dangling down her back, clambering down the side of Keamagawr Bridge at Glengarriff in thrall to the yellow glow of *Hieracium pilosella*, or roaming Droumduff Bog in search of *Hypnum stramineum*. As she prepared to climb into a well near Reendesert in search of *Conferva punctalis*, a man stopped to comment on the danger, and offered to 'fetch whatever it was the good lady had dropped'. His expression while she explained her purpose was a mixture of incredulity and pity. Fortunately most of the places where she collected were uninhabited, at the tip of the world. Only sheep, gulls and the occasional sea eagle witnessed her ramblings. She walked for hours next to mountain marshes without seeing a solitary soul, treading lightly, cautiously, over the boggy ground, eyes fixed downwards for *Hypnum commutatum*. Sheltering from a rain shower under thin trees near Donemark waterfall, her eyes focused on a distortion of the bark that turned out to be *Verrucaria stigmatella*.

Salt marshes with the water pimpernel, *Samolus valerandi*. Glengarriff Woods and the strawberry tree, *Arbutus unedo*. Wet, heathy, mountainous places rich in mosses such as *Grimmia pusilla* and *Grimmia acuta*; any number of old ruins providing sheltered habitation for *Barbula rigida*. She searched out Mr Hooker's beloved *Jungermannia* around Currikimade, a wind-blown, forsaken cliff stretched perilously over rocks and a crashing sea, and found *Jungermannia trichomanis*. At one time, heights made her dizzy, but she'd now grown quite brazen – even foolhardy. Near banks at Coomhola she found *Jungermannia obtusifolia*; above the waterfall at Glengarriff, *Jungermannia trichophylla.*

She went as far as possible on foot, in the gig or by boat, though often forced to stay closer to home, because of weather, or the demands of Jack and her mother. This had its advantages; as she gained ever more intimate knowledge of the place, she saw increasingly beyond the surface to the riches beyond: the gooseberry bushes in the garden at Ballylickey hosting *Orthotrichum striatum*, or *Pterogonium gracile* on the wall below Ballylickey Bridge. An old hawthorn in the top field smothered in *Lepraria ochracea*. In the cove near the boathouse she took account of *Inula helenium*. And buried within the Deep Glen near Ballylickey, the trembling, translucent fern, *Hymenophyllum alatum*.

Illustrating was another way of seeing; even more intimate than looking through the microscope, a process she never failed to think of as invasive, the specimen taken apart to discover its inner workings.

She allowed herself the luxury of two candles, one to the right, one to the left. Their warm glow spread across the table, which she'd covered with an old piece of cloth to protect it from splashes. A flat piece of wood provided a base, over which she laid a sheet of drawing paper, purchased from Edwards on Castle Street in Cork (a treasure trove of papers and ink; she felt a twinge of pleasure whenever she entered its doors). The paint box, of mahogany and mother of pearl, held several cakes of Reeves & Woodyer watercolour paint, four small palette dishes, a large brush, a quill brush, a pen nib, a sponge and a marble palette. Alongside, a mug, filled with water. Two pails on the floor, one to take the dirty water, one for refilling.

She considered her colours. The selection of greens, from bright to dark, were particularly useful. She'd become adept at mixing, and took particular pride in her vivid purples and pinks. The plant before her was a dull brown-green, found in that delicate and miraculous state, full fruit. Her fear: in drying it, she might spoil it. By painting it, she would preserve its freshness. She'd already drawn the outline of the plant, painstakingly copied from the image within her magnifying glass. Now wetting the brush, she dissolved the surface of the paint into colour, and quickly transferred it to a palette dish. She wet the brush again, dabbed it in her mouth and sucked on it, sharpening the tip. There was a faint but definite sweetness under the musty taste of chemicals: honey, added at manufacture so that the solidified colour could be made liquid.

A final dip in the paint. She paused, looked again for a time at the specimen. In advance of touching the paper, the brush should know its business. Then, almost without thinking, a quick assured movement, confident but taking great care, spreading the colour across the drawing, adding life to the dull pencil marks.

She worked for hours, bent over the page, while the house creaked and the wind moaned in the chimney. When finished, she straightened and put down the brush. Her spine screamed; a headache began its slow, stabbing progress. Still, looking at her *Fucus tomentosus*, she was pleased. She'd caught something of its essence, delicacy and complexity. Fading by the second, it had been transplanted to the page, perfect and preserved in pigment and water.

Beyond withering or decay.

FIFTEEN

Her immediate reaction to Dawson Turner's invitation: impossible. They needed her at Ballylickey. Jack would crumble into melancholia, never leave his room. Her mother would fret and pine, might perhaps take ill. There wasn't money to spare for the passage. The garden would dry up and die, without her constant care.

Then tendrils of hope stirred, put out shoots. Jack had managed before she ever came from Dublin. She could write to her mother, every day. Dan would not let the garden die. Was the cost of the fare so prohibitive? She would only stay a month, return home by autumn. Possibility grew, occupied her mind constantly. Images of herself, conversing with Dawson Turner, collecting by the Yare, examining plants together – no doubt he had a far finer microscope than her own – crowded her thoughts, no matter how she tried to suppress them. Several times she opened her mouth to speak of it, but shut it again, swallowed the impulse.

One morning Leonora complained of sore limbs and dizziness. The light hurt her eyes and she asked for the shutters to be closed. She ate little that day except a biscuit and a few spoons of cabbage soup. The next morning when Ellen came to wake her, she could not get out of bed, could scarcely sip the water Ellen held to her lips. The mug chattered against her teeth, water spilling down her chin. She swooned back onto the pillows, her head lolling to the side.

Kate hurried in. 'The master will not stir either, Miss.'

'What do you mean?'

'He says he can't move. It must be the same affliction as the mistress.' Ellen pushed past her and hurried to Jack's room. Opening the door, she whispered, 'Jack?' He lay on his side, facing away, bedclothes knotted around his legs. She came around and leaned over him. His hair was gummed to his forehead with sweat.

'Jack.' She shook his shoulder. 'Jack!' A low grunt. 'What ails you?' she said. 'Are you ill?'

He opened his eyes, slowly, as though the action pained him. 'I'm thirsty,' he said. 'My head hurts. And my back.'

She filled a glass beside his bed; awkwardly she raised his head. He choked, coughed, swallowed. Rearranging his bed-clothes, she said, 'Stay in bed. You have a fever. It will pass.' He didn't reply but slumped back into sleep.

On the landing, Kate asked, 'Should we send to Bandon for the physician?'

She thought. 'No. He might prescribe something that would frighten Mother. And Jack distrusts all physicians. If their conditions worsen, I will reconsider.'

For two weeks she shuttled between their rooms. Her mother gave most concern. Her skin flaked and cracked, brittle as aged manuscript. When Ellen put her ear to her chest, the beating of her heart seemed faint and feeble. Seized by terror, she gave in and sent for Dr Downey. He arrived the following day, dressed in black, and spent half an hour tapping, prodding, poking while he grunted and shook his head; finally he prescribed thin broth, hot compresses, and a course of foul-smelling yellow pills Ellen had to mash and feed into Leonora's mouth. He wrote out a bill for five shillings – an exorbitant amount – in an indecipherable hand and left them to it.

She and Kate took away the sweat-soaked linen each morning and set it to boil over the fire. Leonora moaned as they gently lowered the clean nightdress over her head and manoeuvred in her stick-like arms. Kate had strained her back lifting an inert and feeble Jack onto the chamber pot; now she hobbled about, her face a grim mask of discomfort. Ellen could not spare her. She herself scarcely made it downstairs, or had time to eat anything other than a slice of bread, drink a mouthful of milk.

Late one night while she sat with Jack, dozing, he shouted her name. She got up, went to the bed. His eyes bulged, wide and glassy.

'What is it, dearest? Do you want a drink of water?' He nodded and lifted his head slightly off the pillow. She held the glass to his lips. Sinking back on to the pillow he said again, 'Ellen.'

'I'm here,' she said.

'I have never said it.'

'Said what?'

His breath sounded shallow. 'I'm glad you're here.'

'I'm very glad to sit with you, any time you like.'

'Here, in Ballylickey, I mean.' She didn't reply. Fervently, he said, 'I'm glad you came back.'

'Shush.'

'I want to give you something.'

'It can wait.'

'No. I want you to have it now. Go.' He wheezed and stopped. She waited. 'Go to the bureau. Open the bottom drawer.'

It resisted her pull, stiff as if infrequently opened. She smelled damp. It had been stuffed with linen, much of it evidently unwashed. 'Towards the back,' he said. She groped past the tangled heap of shirts, cravats, nightshirts.

'Whose clothes are these, Jack?'

'Emanuel's, I suppose, from years ago.'

'They should be washed, mended. It may be possible to use them.'

'Never mind that now,' he said. 'There should be a box in there, somewhere.'

She touched a small, hard object, pulled it out, showed him. 'Is this what you mean?'

'Yes. Open it.'

A locket, the size of a man's thumb, heart shaped with a clear face of rock crystal, rimmed in gold. Behind the face, a miniature in sepia ink, of a girl kneeling before a tomb. She appeared to be writing something on the stone.

'It was Katherine's,' Jack said. 'It's been in that drawer for close to twenty years.' Ellen studied the miniature, turning it around in the poor light.

'There's an inscription. I can't make it out.'

'She had a sweetheart, they say. He died.' He swallowed. 'A pathetic tale. She painted the miniature herself. Like you, she had talent. You've seen the silhouette she did of Father, and her own portrait. Mother must know of the locket, though she has never spoken of it. To me, at any rate.' There was silence for a time. One of the tallow candles burnt out with a hiss and a rank whiff of spoiled meat.

'Jack?' No answer. 'Jack?'

He coughed. 'What?'

'Why was it put away like that? In the drawer?'

'This was Emanuel's room, you know, until he left for London. I asked Arthur about it. He said the locket was Katherine's, that Emanuel put it away after her death, and not to speak of it again. It must have pained him, to see it. He admired her. As mad about horses as he is himself, and almost as good a rider.' He spoke with effort, the tendons of his jaw rigid. 'You must have it. You must! It's fitting.'

'Lie down. Rest.' She put the locket back in the box, snapped the lid and slipped it inside her skirt pocket.

In the cold, exposing light of morning, she read the inscription. *'C'est tout ce qui m'en reste'*.

This is all I have left.

※

Care of her mother and brother consumed every waking moment, invaded her half-sleep.

The timing was uncanny. Almost as if they'd read her thoughts, knew of Dawson Turner's invitation to Yarmouth.

She told herself: stop. They depended on her utterly. They could not be blamed for their illness, or weakness. Still, in some disorderly part of her mind, the thought persisted, like a parasite, so that when she succumbed to the fever, she almost welcomed it. No need to carry guilt, if the body took on its own censure. Headache that blinded her behind a red mask. Nausea that left her coiled on the bed. Even breathing brought pain.

'Ellen.' Her mother's voice, and a tender pressure on her shoulder like the lighting of a bird.

'Mother.'

'Quiet, child. Rest, sleep …' Ellen sank, oblivious to anything beyond her bed sheets. Summer passed outside the window. Flitting cloud and sky, blue, grey, purple, black. Raindrops, bees, bluebottles dashed against the glass.

She was lost: too ill to bemoan missed days of collecting and drawing, letters unwritten, unsent. Even with her sheets kicked to the floor, the window opened wide as it could go, her skin burned. She clawed at her hair, pulled away handfuls. She heard dim mutterings, perceived in shadowed flashes Kate's grim face, Joanna's bulging eyes. She was barely conscious of being propped up, a cold blade against her neck. When she woke the first morning without fever she lifted her head from the pillow with a wobbling weightlessness. She raised her hand and slowly explored the blunt, feathery locks. Most of her hair had been hacked away.

Hot tears ran down her face; she hadn't the strength to stop them. 'Never mind, Miss,' Kate said. 'It will grow back. We had to give you relief from the fever's heat.'

'It's childish of me, I know. After all, it only ever got in the way.'

'It was your best feature, Miss. Sometimes the body knows the loss of something before the mind does.'

'Did you keep it?' She meant the cut swathe of hair.

'No.' Deftly, roughly, Kate pushed up the sleeve of Ellen's nightdress, sponged her arm with the damp cloth. 'We had to burn it, in case it spread the sickness.'

Kate reported that Leonora had made it downstairs for breakfast. Jack had called for a newspaper and ordered them to 'get the damned dog out of his room'.

They were over the worst.

She forced her legs over the side of the bed, took a few faltering steps. A newborn calf, wobbling its way into the world.

'You never felt ill, Kate?' she asked.

'No, Miss Ellen, thank God. Or I don't know who would have kept the place going.'

'Will you help me downstairs? I long to see the garden, at least.'

'Take my arm.' Kate smelled of fresh sweat, though her skin shone clean.

Outside, Ellen screwed up her eyes against the glare. Her heart raced as her body, blood and lungs adjusted to movement. Here were the chickens, shucking their feathers, scratching in the gravel. How full of energy, how clucking, strutting, essentially alive. Through the gate into the garden, brushing away flies: 'Ah! Dan kept it watered.' The last of the cabbage and lettuce, straggling but still leafy green, cupping water from their last dousing. Rows of outgrown onion spires, topped with purple blooms, poked into the sky.

On closer inspection, some of the lettuce had spoiled, rotted underneath. 'Pick what we don't need and give it away to whomever you like,' Ellen said. 'We can begin pickling the cabbage.' She plucked a gooseberry. Over-ripe, starting to soften. Clusters of fruit sagged on the bush, wasps buzzing angrily amidst the droppings. 'And we must start on the preserves.' She took in a great breath of moist air. The sky, perfectly still and windless, was the colour of fine ash.

'I think I'll walk as far as the waterfall.'

'It's too soon for that, Miss Ellen. You're still weak.'

'I want to move my limbs.'

'We'll see how you are in a few days. The waterfall will still be there.'

Despite Kate's protests, Ellen got up early and helped wash and scald the glass jars. They were lined up now on the shelf, glinting in the early light.

The window had been shut against the flies. Joanna's face was already puce from the effort of keeping the fire stoked. She stood ready with a great wooden spoon to stir away the sediment that formed as the fruit boiled. Later, while the women waited for the preserve to cool, they drank tea companionably at the long kitchen table.

'Sickly sweet, that smell,' Kate said. 'Not that I don't love a spoon of gooseberry jam. But that smell ...' Her nose wrinkled. 'Too much, like I'm drowning in it.'

'On the bleak days of winter, it will remind us that summer must return. Warmth and bounty, all that's good. A taste of hope.' Unable to keep the emotion from her voice, Ellen bent her head.

Kate drew back her chin, frowned. 'You are not right yet, Miss. Whatever you might say.'

Someone watched her, eyes boring into her back. She spun around, looked up. Light glittered off the warped glass of the windows. Just then Kate emerged through the back door, hands on hips, in her blue blouse and skirt, white cap. Ellen waved. *Don't worry. All is well.*

Into the woods, the perpetual twilight. For half an hour she picked her way over tree roots and stray branches, through ferns, nettles, late celandine in stray patches of sunlight. As she approached, she heard the first rushing amidst the trees. When she emerged into the clearing, the full power of it thundered into her ears.

She'd pined for this place, even more than the shore.

The spray landed on her face and her nostrils filled with the clean rush of moving water. She sat on a fallen tree trunk and idly picked at the rotting wood, setting loose a nest of woodlice across the moss-covered surface. At first glance the moss seemed amorphous; looking more closely she observed its flagelliform branchlets. *Dicranum flagellare.* A miniature world of creation, death, renewal … when she lifted her head again her vision blackened, pulsing with stars. She steadied herself, one hand on the trunk, waited for it to clear. Then she stood. Slowly the trees came back into focus, as the blood settled within her skull.

She needed to move. She would walk some of the way towards Glengarriff.

Out of the woods: the path grew steep and she climbed slowly, sweating, to the foot of the mountain, above the waterfall and along the banks of the river that fed it. She passed a dank cavern, down which trickled a clear shining thread of spring water. A clump of deep, almost blackish green within caught her eye. She stepped closer, bent over, picked out the individual plant and held it to the meagre light: tiny, barely one eighth of an inch across, its leaves edged with spiniform teeth. Groping in her pocket – blessed luck – she found her trowel, buried there from a previous outing. She dug out a fair sample, then stood and brushed off her skirt. Time to go back, before Kate sent out a search party and Dan came thrashing through the trees to get her. She hurried, almost slipping on a wet rock, back towards the house.

Under the microscope the plant showed its spherical seeds, covered with opaque dots, probably minute tubercles. Her heart quickened. New treasure. How pleased Mr Hooker would be.

Jack still wouldn't come to church. Each time Leonora asked him to join them – she fretted for his soul – he mocked her. 'Women's nonsense,' he said. 'I have better to do than listen to children's morality tales.'

In the carriage on the way to Bantry, Ellen said to Leonora, 'You know he hasn't left Ballylickey for almost a year.'

'Is it really so long?' Leonora shrugged. 'He is proud. I would not force him. At least he seems happier now you're here. More stable.'

'Stable?' The previous week he'd thrown a glass against the wall.

During the sermon she cleared her mind of him and tried to concentrate on the vicar's drone. A shiver trickled down her neck. The church was cold all year round, icy in winter. The thick stone allowed little sunlight to penetrate. Candles were lit either side of the altar, their flames steady and fierce. On the chair next to her, Leonora muttered her prayers, utterly absorbed in some inner dialogue with God. Ellen hadn't yet found this kind of solace in prayer, and wondered if she ever would.

A tickling in her throat: she swallowed hard to suppress it, coughed quietly into her fist. No use. The spasm clutched at her chest, overcame her. She fumbled for her handkerchief and bent double, coughing into it for a minute or more. At last it eased. She stood upright, wiping her eyes. Heads had turned. Her sides ached. She folded away the stained handkerchief, nodded to Leonora – *it has passed* – and joined in the final prayer, even as her voice cracked.

As soon as service ended, a hum of conversation rose. For many of the women this was the only opportunity for society in the week. Ellen and Leonora filed with the others into the sunshine and stood for a few minutes, greeting acquaintances. Reverend Smith approached, extending his arms as if to embrace them. 'Miss Hutchins. Mrs Hutchins. We're glad to see you.'

'Reverend. A most enjoyable service,' Leonora said.

He nodded, receiving his due. His pepper-grey hair fuzzed about his head, like a dandelion gone to seed. He gazed into the sky and sniffed.

'A beautiful day,' he said. 'Though threatening rain. With luck I'll be home by the time it arrives.' He nodded towards his horse where it grazed at the grass verge, a handsome, shining beast donated by Lord Bantry. 'I have miles more to travel, three more services to conduct before evening.' He craned his neck at Ellen. 'How long, dear Miss Hutchins, since we last saw you? Months, surely?'

'We have had illness in the house,' Ellen said. 'First my mother and my brother, then myself.'

'I had heard, but hoped not all were afflicted.' His little eyes, more cunning than intelligent, scanned Leonora's face, searching for remnants of sickness – ill colour, sunken cheeks, dull eyeballs – then flicked to Ellen's. She boldly met his stare.

'Thank you for your concern, but my mother is recovered now,' she said. 'Much her old self.'

'And your own health?'

'I am well, thank you.'

'I'm glad to hear it.' His eyes slid away and he coughed delicately into his hand. 'Any news of young Samuel?'

'He is in Trinity now, having finished his schooling in England.'

'Will he come home for the holidays?'

'Perhaps,' Leonora said.

'Well. Soon, we hope. I'll continue to pray for you.' A tight smile on his lean features, he backed away, already nodding at another congregant.

'Even the Reverend noticed your coughing,'Leonora said.

'Dust, caught in my throat. It sounded worse than it was.' Her mother gave her a sideways look but said nothing. They came out onto the road and got into the carriage. Spindly trees filtered shafts of early autumn sunlight through the windows, and they winced in the sporadic glare. Leonora said little on the journey. Ellen leaned her head back and let the movement of the wheels shudder through her body.

That evening she sat at her writing table, by the window facing the sea. She waited, quill in hand, gazing at the shifting colours that changed as the light changed, mirroring the brighter gleam of the sky, reflecting its soft yellows, pinks, purples and silver blues, filtered through cloud. Her brain cleared, she found the necessary words: firm, decisive, rueful, with, she hoped, the right note of regret. She was sorry, she said, but she was bound too closely to home to visit Yarmouth at this time; she hoped Mr Turner might call on her some day on his way to the mountains of Kerry.

It would be nothing to the Turners to have their invitation refused, they who apparently had so many visitors. Disappointment was surely all on her side.

Like a fog that seemed to descend after the letter had gone, it lingered far longer than it should have, made all her endeavours seem muted, colourless.

Another winter of confinement loomed, as if she entered a dark tunnel: short, dark days of mist, long nights of howling gales that hurled slates from the roof and knocked trees. Rare days of clear skies and pallid sunshine, though the ground remained bogged underfoot from the quantities of rain. She measured the hours in spent candles and firewood, swaddling herself in shawls and cowering in the rooms of the house kept warm all day – the kitchen and parlour – like an animal in its den.

Vegetation shrivelled away, the landscape reduced to an outline of itself, as in an etching by a Dutch master in one of Dr Stokes's books. Alone in the bleak garden one December morning, pruning back the last season's growth, she heard a sudden cackling din fill the sky. Geese: a flock of dark arrows shooting towards some winter resting ground. She stood, open-mouthed, shears in hand, until the last of them disappeared from view.

A note arrived from Co. Kildare, wishing all at Ballylickey a pleasant Christmas. Caroline sounded settled, less … *unmoored*; Charlotte, now almost three years old, seemed to act as ballast, grounding her mother as nothing else had: *She is so tangible. Her chubby little arms, her sturdy legs, her personality, which is by turns cheerful and solemn. I believe she is my purpose on Earth … One day I will take her to visit you, in your far-flung corner of Ireland.*

Ellen sent her a small watercolour, a view of the bay and mountains. *This is the view from my writing table. Framed, it may make a pretty display and remind you of your ever affectionate friend, E. Hutchins.*

On Christmas Eve the fourteen-year old son of their neighbour Cronin arrived at the back door, cradling a large bloodied parcel. 'My father butchered the cow, Miss,' he said. 'He sends this with his compliments.' He hoisted the parcel on the kitchen table and wiped his hands on his trouser legs.

Kate unwrapped the meat, exclaiming, 'He's given us the heart.' It was larger than Ellen would have imagined, the size of a man's head, cloaked in a layer of cream-coloured fat.

'Thank your father, John,' Ellen said. 'Kate, bring him some cake.' The cake had been ripening for weeks in the pantry. John, a solid youth with dark hair and chapped fingers, swallowed his slice in three bites and licked the last crumbs from his lips. Then he bowed awkwardly and ran out the back door, his boots clattering across the cobbles.

'We haven't had beef since the spring,' Ellen said.

'The heart needs to be steeped,' Kate said, 'Or the blood overpowers the taste.'

Ellen found it later in the pantry, trimmed and sunk in a bucket of milk, leaking blood pinking the liquid. Though saliva had earlier flooded her mouth, now she looked away, faintly sick.

'Where are you going at this hour, Miss?' Kate said, seeing Ellen pull on her boots.

'Out with the dog to clear my senses.'

She staggered into the kitchen just as the last light seeped from the sky, subsumed by branches of holly and pine, trails of ivy. 'Help me, Kate.'

'I have enough to do, Miss.' But she stood obediently with bunches in her arms as Ellen arranged the holly across the mantel, and climbed on a stool to tie greenery over the lintels.

On Christmas Day, they dressed in their best clothes. Jack wore a clean cravat, Ellen Katherine's locket. Leonora started when she saw her, but said nothing; through the day her eyes strayed to where it nestled against Ellen's breastbone. For dinner they ate the heart, sliced in steaks and seared in the pan, and afterwards, a trifle that Ellen had made, as like the one she'd eaten in Harcourt Street as she could manage. Jack opened a bottle of claret, and Ellen drank a glass. After finishing the wine and winning several games of cards, Jack fell asleep on the sofa. His fingers twitched as he snored.

'Leave him there,' Leonora said, drawing a rug across his legs. 'I'll go to bed. Will you come, Ellen?'

'I might sit with Kate in the kitchen for a while.'

She was drinking tea at the table, and made to stand as Ellen came in. Ellen gestured at her to stay sitting. 'I'm sorry you could not go home like Joanna did.'

'Too far to Ballydehob for one day's holiday. Besides, I'm better off staying. My brother's wife, all those children, in that cramped room ... there's more comfort here.' Joanna's parents shared a small cottage with three of her siblings, almost full-grown, and two ancient grandparents, but Ellen didn't comment. Shadows from the pine bough hung above the fireplace reared across the ceiling. The dying embers still gave off heat. Curled in her basket, Lettie whinnied, chasing rats in her dreams.

Kate lifted her teacup. 'Will I tell your fortune, Miss?'

'That's superstition.'

'Aye. But where's the harm? Besides, I have a gift for it. Didn't I foretell that Joanna would find a sweetheart? The next month she started walking out with Donie Murphy.' She changed her tone. Bossy, wheedling. 'Sure, who is there to see? Wouldn't you like to know what lies ahead?'

'No one can tell that.'

'Isn't Christmas a time for games?' Kate rose and took a second teacup from the shelf, sat and pulled the candle closer. She poured a long stream of golden tea. 'Drink, until you've about a teaspoon left.' Ellen gulped it down as she was told and held out her cup. 'Close your eyes. Ask a question, or for guidance. In silence, mind.'

'I should know better.' But she closed her eyes obediently, made her face passive. Unbidden, an image flashed into her head. She opened her eyes and handed the cup back across the table, shaking her head at her childishness: excusable in young ladies, feigning boldness in their fussy boudoirs, or in ignorant farm girls, giggling around tables like this one once their father had safely gone to bed.

Kate swirled the cup counter-clockwise. Her forehead creased in concentration. She flipped the cup down in the saucer and let the liquid drain away. Then, more delicately, she inverted it the right way up. With a clank the last log collapsed against the grate. The dog jerked her head and Ellen's heart leaped. Kate didn't move.

'What do you see?' Ellen asked. Kate frowned and bit her lip. Ridiculous, Ellen reminded herself, the entire performance. 'Tell me, then we can go to our beds.'

'A stranger. More than one,' Kate said. 'And you'll reunite with a loved one.'

'Who?'

'I can't say,' Kate said slowly. 'But it will bring you joy.'

'Well, that's pleasant enough.' She leaned forward, intrigued in spite of herself. 'Anything else?'

Kate had lifted the cup again and stared into it. A full minute passed. Embers hissed in the grate. She shifted in her chair, as though in discomfort. A strange look came across her face: confusion chasing fear, chasing sorrow. Then her eyes went empty and her mouth hardened.

'What is it?'

She clinked the cup down in its saucer, splashing tea on the table. 'That's it. That's all I see.'

'Truly?'

'The leaves aren't clear.'

'You saw something. Is it about me?' Kate paused, then shook her head. 'Would you tell me if it was?'

'What kind of question is that, Miss?' She sounded angry, as she often did when moved, a way of defusing the deeper feeling.

'No matter.' The lines around Kate's mouth had deepened; Ellen noticed hollow shadows under her eyes. 'It's been a long day, if a festive one. You look exhausted, Kate. Will you not go to bed?'

'I will go, gladly, once I've tidied here.'

'Let me help.'

'No, Miss.' Her voice was firm.

'I need a candle—'

Kate got to her feet. 'I'll get it.' She took a fresh one from the drawer and fitted it to a sconce, touched it alight from the candle on the table.

'Goodnight, then, Kate.'

'Goodnight, Miss.' She bobbed a half curtsy, then sat again, heavily.

Ellen went to the door, paused. 'Kate? Has aught upset you?'

Her back was hunched, her hands clasped. 'No, Miss.' After a pause, 'T'was foolishness, like you said.'

'I take it lightly, I promise you.'

Kate nodded but still stared ahead, fixed on something unseen.

Climbing the stairs into the dark, holding the candle flame ahead of her, Ellen wondered: what had she done, said, to offend? Her heart chilled at even the possibility of ever losing Kate.

SIXTEEN

The tone of the letters between Mr Turner and herself had changed. He poured out his thoughts freely, gave more details of his life. Children, wife, friends who had died.

She cherished these glimpses of his private life, stitching them together to assemble a picture of his world. In Ballylickey, banal fractures indicated that time proceeded, unstopping. Joanna would marry that summer. The butter spoiled, the bucket on the upstairs landing filled after a night of heavy rain ... the sun rose, set, rose, set, though some days it remained lost behind heavy cloud. She nursed a headache. Bled, or not.

Stagnancy. Offset by the life under her microscope.

They were again between winter and spring. Plants lay dormant and she had no reason to write. At night, and during the dull tasks of day, she harboured dark thoughts. Her mind had always found opportunity to turn on itself.

Much of what she and Mr Turner wrote about concerned that which they could otherwise not discuss. Seeds, male and female parts, fertilisation – Dr Stokes had once explained that the classifications of Carl Linnaeus had been initially deemed unsuitable for women, so literal was his terminology. She read some of Linnaeus's more 'flowery' writings and considered them almost naïve in their romanticism. Living on a farm, she couldn't help but witness how procreation occurred amongst animals. She presumed it to be little different between man and woman. A practical means to an end, a matter of transference.

Romance seemed something quite other. Pleasure, within. The longing to touch and be touched. Pressure on her back, fastening buttons. Creeping feet, seeking warmth. Water trickling from the washcloth down her thighs. Fingers, brushing against hers, flinching away as though burned.

A woman who did not bear fruit, was she not still a woman? Had she then no need for love?

She should have liked a child. Without a husband, the means of transference, produce seed within herself, self-fertilise. Create new life, utterly contained, self-sufficient.

How shocked Mr Turner would be, were he to hear these fantasies. Perhaps he'd laugh heartily, thinking her scandalous or simple-minded. Send for the physician's orderlies in their long coats, to come rattling their chains.

LONDON AND YARMOUTH,
ENGLAND
MAY 1809

SEVENTEEN

The coarse-looking fellow in the seat opposite belched, grunted an incoherent apology, belched again, and honked his nose loudly into a grubby handkerchief.

For the tenth time that morning, Dawson cursed the journey. He detested London, hated leaving Mary at this time. Unfortunately it was necessary. The forgery that had come to light in the bank meant it was essential to meet with the lawyers as soon as possible and halt the disaster. He took out his own handkerchief, crisply laundered and smelling of lavender, and held it to his face. A similar scent at the hollow of Mary's throat. He closed his eyes. The wheels hit a rut in the road and his neighbour lurched heavily on top of him. Confusion, as he awkwardly righted himself and resettled in his seat. Now his thigh pressed against Dawson's, an imposition to endure for the next hundred miles.

The timing couldn't have been worse, with another new baby so soon after Hannah's birth the previous year, this time a boy named Dawson for his father as well as his dead brother, arrived only a week hence. Mary had said, smiling, 'We'll be perfectly well until your return,' and continued gazing into the child's face, as though silently communicating to the little creature other thoughts, deeper, more intimate.

Dawson would not have left them unless absolutely necessary.

He tucked away his handkerchief, shifted on the seat. Something else contributed to his disgruntlement: William Hooker, chief cohort amongst his botanist friends, had impulsively decided to decamp to Iceland, as he'd been threatening to do for months. Passage had become available; the weather and conditions were favourable. Dawson would miss his company, but envy had also taken root like an insidious weed. A pleasurable month's botanising in a country of legendary alien and mysterious landscape, culture and people, accompanied by the most amenable and intelligent of companions – he, Dawson, could only dream of such an adventure. Instead he must give way to the double binds of family and work. Of course Hooker was of independent means and unmarried.

Dust swirled in noxious clouds and pervaded every window. Nursing his mild asthma, Dawson slept with the sash closed, sweltering under his sheets, for the nights were unseasonably warm. During the day, between tedious meetings with his lawyers, he met botanist acquaintances at the Linnaean Society at 9 Gerrard Street and discussed the publication of *The Synopsis of British Convervae.*

Eventually, when his nerves had become frazzled almost to breaking point, a blessedly cool drizzle descended on the city. Exiting an appointment, he decided to have a quick turn about the Watercolour Society exhibition. He ordered a hackney carriage and directed it to Spring Gardens.

Walking through the airy rooms, he wished Mary could have come with him. He stopped before several paintings, allowing himself to imagine them on the walls of Bank House. The exhibition was quiet at this hour, echoing with the agreeable hush of contemplation; only the sound of his clicking boot heels and the occasional tap of his cane against the wooden floors. Portraits of stately gentlemen, pretty women, churchyards, fens, mountains, ships and seascapes, bowls of fruit and baskets of fowl: he scanned them all quickly, appraisingly, all the while keeping an eye out for Thomas Heaphy, a painter he particularly admired. Ah! Here he was, hung between Henderson and Varley … two paintings, each of a vernacular scene. The usual colours Heaphy favoured – gold, cream, pale blue, the contrasting glow of rose pink or sapphire. Here a family assembled in a cottage kitchen, the husband seated, receiving treatment of some sort to his injured leg by an old woman, his wife's breast exposed to the suckling mouth of the baby that nestled on her hip. Dawson admired the details of humble kitchen utensils, exalted to objects of beauty – a saucepan, a striped mug.

He moved on to the next: a fish market, in all its raw vitality. The bright centre of the painting immediately drew his eye, as the artist had surely intended. He bent closer. A young gentlewoman, dressed in white, with a gold locket around her neck, raised her hands to refuse the fish proffered to her by a swaggering fishmonger, while an officer's boy attempted to press a letter on her blushing servant. In the background: sailboats, a row of tumbling cottages. Haggling and quarrelling, thick-armed fish wives, hands squarely on hips, in full command of the jostling crowds. Dawson had played around Yarmouth fish market as a boy, amidst the knives, the stink of guts, the wandering tug of pickpockets; such scenes were as familiar to him as the comforts of his drawing room.

His eyes flickered back to the young woman in white. Heaphy had added a flush to her cheek and painted her hair, just visible under her straw bonnet, a shining gold. The ends of her shawl draped over her upheld arms, giving her the look of a graceful yet defiant angel.

Eventually he checked himself, pulled out his pocket watch. He had spent almost an hour here already. At the desk, he rang the bell. The secretary emerged from an anteroom, expression ready for servility.

'Yes, sir?'

'I want to enquire about a painting.'

'Certainly, sir.' The secretary pulled a large ledger from under his desk.

'Thomas Heaphy,' Dawson said. '*The Fish Market.*'

The man ran his finger down the pages. 'I regret that painting is sold.'

'May I ask for how much?'

'Four hundred guineas.'

Dawson drummed his fingers on the countertop. 'Have you an address for the purchaser?'

'I can't give out that information, sir.'

Dawson took a coin from his pocket, placed it on the counter. The secretary cleared his throat, placed his hand over it, slipped it out of sight.

'One moment, please.' He wrote the address out in a careful hand. Dawson glanced at it before slipping it in his pocket.

'Thank you. Good day.'

'Good day, sir.'

The drizzle had eased, the afternoon now damp and muggy. From his inn Dawson wrote a note offering the sum of five hundred guineas for *The Fish Market* – if accepted, he could arrange to have the painting collected at the owner's convenience. He rang for a servant and ordered the note to be delivered.

As he dressed for dinner later, there was a knock at his door: a servant with a note. He cut the seal and shook it loose. Quality paper, crackling as it unfolded.

I am not prepared to sell the painting for the specified sum, and indeed, were you to increase your offer, I would likely refuse that also. I am aware that Mr Heaphy has painted similar scenes and can only advise that you contact him through the Watercolour Society if you want to purchase one directly. One fish market is, after all, very much like the next.

However, I find that I cannot bring myself to part with this particular one …

Strangely frustrated, he tore the note in quarters, left the scraps in the grate, and dressed for his dinner engagement. He had to wait some minutes for his hackney to arrive. It would have been madness to spend such a sum on a single watercolour. He should be thankful he'd been spared from his folly. Had the owner agreed to sell him the painting, he would be no doubt be sitting in the very same spot at this very same moment cursing his impetuosity. Still, the sour feeling lingered right up to his arrival at his friend's townhouse on Brook Street. Hurrying in to the extravagantly lit hallway, escaping the deluge, he made an effort to shrug it off. After several glasses of strong claret he succeeded.

In the morning, a letter called him back to Yarmouth: sickness had entered his house.

❧

He walked into a morbid quiet. Rachel met him in the hall, her eyes lowered. The other servants he met shrank against the wall as he passed.

Usually the children came running in a pack, calling for him. On the first floor landing he found Maria, sitting alone on the stairs. She said nothing, looked up at him through eyes like luminous pools. He pressed a hand on her hair for a moment and continued upstairs to the second floor.

Mary was in her bath, arms wrapped around her knees. Her thick hair fell around her shoulders, damp at the ends and clinging to her skin. 'I was covered in soot,' she said. 'We've been burning the linen and furniture all day.' She didn't look at him. The steam from the water smelled of almonds.

'What happened?'

'*Cholera morbus*, the physician said. We'd only just sent for you when …' She swallowed. 'When he deteriorated. He was too weak, from the vomiting and diarrhoea.'

Dawson sat down heavily on a chair, weary beyond words. 'Mary …' he said.

'I said we should have given him another name. Having the same as the little one we lost invited bad fortune.'

'That had nothing to do with his dying.'

She stared ahead. 'Even so, he could at least have had his own.'

'Next time ...' he said.

'Pray do not speak to me of a next time,' she said. 'Not now.'

Night after night, he stayed up late in his study, waiting for her to fall asleep. He couldn't bear to lie beside her, sensing that her raw eyes stared into the dark, fearful of moving a leg or arm and showing that she kept him awake.

He opened the book, as he had done already, countless times, at Canto I:

> As journeying midway on the road of life,
> Bewilder'd in a dusky wood, aghast!
> I stray'd, far-devious from the path direct.
> It were a hard attempt to say how wild,
> How savage, and how vast, that forest rose,
> For even yet remembrance on my mind,
> Bitter as death, renews the pale affright.
> But to display the good, I must recount
> All other wonders there disclos'd to view.
> I know not well, how thro' the horrid gloom
> I enter'd; for dull languor fill'd my soul
> What time I wandered from the certain way.

Yes, he had been here before, but familiarity did not make the road easier. If anything, the dread increased. There was comfort, however, in having Dante to tread with, if only for the hour he spent under the glow of the oil lamp.

They each stood on an island, separated by a dark sea. He tried to breach the divide, but Mary remained stubborn in her sorrow. He coaxed her to eat; she pushed her plate away as though repelled. He entreated her to take a walk; she lay on the sofa, feigning sleep. Not yet recovered from the strain of childbirth before the calamity, her cheeks grew hollow, the bones of her wrist prominent under her cuffs. Walking about the house she stumbled, gripped the back of a chair, the banister, before moving slowly on, waving away the hands that reached to help her.

In moments of exhaustion he fought to suppress a spike of resentment. Did he not have his own cares, the responsibility of the bank, as well as the commitments he had made to his botanical studies, his colleagues? She acted as if he were immune to grief, as if she had greater entitlement. He reminded himself how she had carried the child inside her for those long months. An image came to him, of her stitching at a palm-sized white bonnet, her face puckered in concentration.

He brushed at his eyes, breathed deeply, shook his head and freed himself of the memory. Moving to his desk he opened a drawer and drew out a sheet of writing paper. He took up his quill.

Dear Miss Hutchins.

Was *she* the bright angel amidst the mundane? She must be practical, of the earth, literally, demonstrated by the sheer quantity of plant samples she continued to send. No delicate flower, she who grubbed around beaches, in tree roots, rock pools. Yet the poetry in her descriptions of plants elevated her beyond a mere forager. The heated prose of her letters suggested a woman of passion, of exceptional fervour and intellect. Surely her physical form would reflect that nature, in beauty and grace ...

He'd last indulged in this type of reverie as a younger man, before his marriage. Truly, he was not himself.

❧

Striding towards him, hand outstretched, his friend Joseph Woods.

'I didn't know you were visiting Sir James.' Sir James Edward Smith, president of the Linnaean Society, raised his hand in acknowledgement from across the room.

'A happy coincidence, for I hadn't known myself until a short time ago that I could come.'

'How is your family? Growing as fast as ever?'

'Very well. That is, as well as can be expected.'

Woods reddened. 'That is, I mean to say ... I heard ... forgive my stupidity, Turner.' His eyes popped in alarm from his bony skull. A lanky man with a harried expression, not overly blessed with social graces beyond those of his birth and education.

Dawson shook his head. 'No matter, Woods. You meant no harm.'

'Still ... it was thoughtless.'

Dawson changed the subject. 'How goes it? You have a new commission?'

'Several.' Woods was an architect, establishing his career.

'Have you had time for botany?'

Woods frowned and pulled at his nose, a habit signifying perturbation. 'Well, I do my best, when I can. The demands of my profession ...' He trailed off, shrugged. Dawson understood. Who better?

Woods brightened suddenly. 'I wrote you of my Ireland trip.'

'You still intend to go?'

'Certainly.'

'I envy you.'

'And now that I see you,' Woods said, 'it reminds me – I thought I might call on your lady botanist.'

Dawson blinked. 'You mean Miss Hutchins?'

'The lady from Bantry. Your esteemed plant collector.'

Dawson had mentioned Miss Hutchins to Woods as a possible contributor to his own studies. 'I didn't know you would be anywhere near those parts.'

'I believe we pass quite close by, travelling from County Waterford to County Kerry. It seems a good opportunity to meet the lady and see some of that locality. Could you write me a letter of introduction?'

'The lady lives in some isolation, with an elderly mother and crippled brother. It may not be appropriate. Besides, having never made her acquaintance in person, it's hardly my place to vouch for others.'

'I thought there was considerable intimacy between you as correspondents of long standing.'

'As you know, the intimacy that develops on the page is often inflated. It doesn't negate the need for the usual formalities. In this instance, the fact remains that I haven't actually met Miss Hutchins.'

Woods listened intently, his head to one side. 'Still, a word from you …'

Dawson let this hang. 'Do Leach and Dillwyn still talk of accompanying you?'

'If their schedules allow.' Leach the beetle collector and Dillwyn the botanist: agreeable gentlemen, yes, but obsessives, Dillwyn tending to pomposity. 'Since your Miss Hutchins provided illustrations for Dillwyn's *Conserve* he is most anxious to meet her.'

'I wish you would not speak of Miss Hutchins as *mine*, Woods. It hardly seems appropriate.' Yet how to explain the proprietorial surge of irritation he felt at the prospect of his good friends being received in Ballylickey House?

Woods had visibly shrunk at the rebuke. Dawson immediately softened. 'Let me think on it,' he said. 'I'll mention it when next I write and see how she responds.'

'I dare say she would receive us anyway, as English travellers. Irish hospitality on these occasions is renowned. You know I have Irish blood myself, through my mother.'

'I had forgotten that.'

'There's no chance of you joining us, Turner?'

'I must wait for another year, another season.'

A few days later, he finished the letter to Miss Hutchins, cautiously recommending Woods, though stopping short of an introduction, for now. He shouldn't resent his young friend's freedom; he had knowingly made his own choices. And such estimable guests would surely divert the lady. Had he not fretted over her isolation?

He twisted in his chair and stared at the wall for a moment, thinking. Then he turned back to the letter and scribbled a footnote, mentioning Heaphy's painting of the fish market and the woman in the white dress.

Once the letter had been sealed and sent, this action confounded him. Why should he tell of such a meaningless episode, particularly when there was no chance she would ever see the painting herself? (The only thing, he thought, more tiresome than listening to an acquaintance recount a dream, is to listen to them enthuse over an artwork that you yourself were never likely to see).

For some reason, he had felt compelled to tell her. Another connection, perhaps. So much separated them, beyond the Irish Sea.

BALLYLICKEY, CORK AND DUBLIN
JUNE 1809

EIGHTEEN

According to Dawson Turner, Mr Hooker had at last embarked on his Icelandic voyage. He intended publishing a journal of his travels.

What did she know of Iceland? Nothing, beyond her own imaginings. A blue-white frozen world. A blank vista. Where a human could only survive by burrowing themselves inside their dwellings, though what could these resemble? Surely not like the dwellings of the Irish poor, those roughly constructed huts of mud and straw, with holes cut in the roof and sides for chimney and window ... inadequate for the Irish winter, never mind a frozen tundra many degrees below freezing.

Rivers, mountains, lakes, bogland ... did they shimmer under ice? What type of plant life could be sustained in such an environment? How was any life sustained? What did the people eat? How did they not freeze to death? How did they move from place to place?

Weather and the turn of the seasons defined her existence, and that of everyone, everything around her. Cold snaps that pinched the life from animals, babies, old people. Gales that bent trees double and whipped away haystacks. Damp that fostered swathes of mould on interior walls and infested chests with racking coughs. Rain in every variety: tender drizzle that masked the landscape for days, obliterating the sky; cloudbursts that ravaged picnics and washed away newly planted gardens; storms that pummelled stone bridges and dragged fishermen from rocks. Surface destruction. Then, from the soaked earth, a reservoir of life.

When there are sparks on the kettle and the pot, it's a sign of rain.

Ring around the moon, prepare for a storm.

Did the Icelanders study the sky, the movements of the birds, the antics of animals, in an attempt to read the weather? Had they a hundred superstitions, prayers and blessings to circumvent its mysteries?

All this Mr Hooker would learn on his voyage. If she couldn't tread the landscape (she imagined herself muffled and swathed in animal furs, on some sort of sled pulled by strange, horned creatures), she would be one of the first to read of his experiences. Dawson Turner had promised it would be so, that she would receive a copy from Mr Hooker himself. What her collecting, illustrating, drying and dispatching of plants across Europe had earned her: the willingness of great men to share their knowledge with her, as an equal.

No word from Tom Taylor in months. His letters had become sparser the last year. When last he wrote, he had begun studying for an MA in medicine. Difficult to leave Dublin, though *he often thought of Cork and wished he could be at Ballylickey once more.* He wrote of botany, from one collector to another, dry observations, cool speculations. He seemed distracted, measuring out his words. What had caused this new reserve? She reread old letters for deeper meaning than what appeared on the surface.

I may settle in Dublin, for reasons of work, though my heart still pulls towards the southwest. If you see a fair specimen of Dicranum fuscescens, *pray save me a sample. That is, if you think of me. I'll collect it when next I visit. Or come to Dublin, bring it then.*

Could he be ill? Perhaps he'd gone to India to visit his father. Or had she mistakenly, clumsily, said something to cause him offence?

Unless they intended going to town themselves, every morning except Sunday a lad went to Bantry for the post. As always, Kate had laid the letters on the dining room table beside the teapot. Ellen recognised his writing and took up the letter quickly. She slipped it into her pocket.

'Who is that from?' Leonora asked.

'Thomas Taylor.'

'Will you not open it? Perhaps he intends on paying us a visit.'

'I must feed the hens.' She scraped back her chair.

'Let Kate do that,' Leonora called after her. Ellen pretended not to hear. In the back hall she took her apron from its hook. Leonora's face had crumpled in distaste the first time she wore it.

A lady doesn't toil like a scullery maid. Too much to be done, Ellen replied. Not enough hands to do it.

Besides, this was light work, hardly like scrubbing floors. The chickens ran towards her on their pink-stick-legs, shrieking with excitement. Reaching into the feed bucket for a handful of grain, she could feel the letter move against her thigh. She sidestepped the distracted birds and stoop-walked into the hen shed. Squatting, she groped amongst the straw, searching, fumbling, now closing her fingers surely, carefully on each egg, filling her basket with a perfect dozen. 'Well done, well done,' she muttered soothingly. At the back door, she rinsed the eggs clean in a bucket of water, then left them aside in the cool of the pantry, separate from the previous day's hoard. She quickly counted: twenty-two, and more every day. They'd soon have to bake. Automatically, she checked along the shelves, the jars of pickled beets and cabbage – running low this time of year – strings of smoked herring and hung beef. A sack of flour, another of oats, the locked larder with sugar, salt and coffee. She sniffed at the milk jug: yesterday's, already on the turn. They'd have to start leaving it in the well bucket to cool. In this weather, it was impossible to keep anything fresh from spoiling. The ham pie, more than half of which was left, would have to suffice for dinner and supper. She hung the apron back on its hook, told Kate, 'I'm going to rest in my room for half an hour.'

'Will I call you?'

'No. I'll come down.'

In her room, she pulled up the sash. Nine o'clock: already the day promised heat. The sea was a powdery blue. A chaffinch per-eeped in a shrub nearby. A rare, God-given morning.

The letter filled one side of one sheet.

He was well. He had met the renowned botanist John Templeton, and accompanied him to Luttrellstown Woods where they found the rare, peculiar, Bird's-nest Orchid. He passed on Mr Mackay's regards. There had been an outbreak of cholera in the workhouse on James's Street, though thankfully it had been contained. If they had a hot summer, it would no doubt resurface. He would spend some time in Dr Stokes's country house in Ballinteer in the Dublin hills; get away from the stink of the city.

At the end, five colourless lines that communicated the most profound change in his circumstances. She read, reread, as though observing a line of rocks between boat and shore, comprehending yet not fully believing in their import or danger. The walls of the sunny room seemed to stretch away, then rush in upon her, so that she found it hard to breathe.

I have some pleasant, if unexpected, news. I've married my cousin Emma Taylor. We had been engaged for some months. For the present I've taken a house on Baggot Street, not far from Mackay. I don't know when next I will come through Ballylickey, if at all, though be assured I will call to you if I do so.

He signed it, *Your friend, Thomas Taylor.*

Downstairs in the parlour, Leonora sat at the table, arranging silver cutlery in rows. 'What a shame we don't have the
opportunity to use this any more,' she said as Ellen came in.
She held out a fork, catching a thin beam of light from the
window. Ellen picked a spoon out of its row. It reflected her
image as a miniature, blurred distortion.

'This was mine, in school,' she said. It had been put away
carefully with the rest.

'You know it amongst the others?'

'I would know it anywhere.' She tapped it lightly in her
palm. 'By these scratches.'

'Well, what news from Mr Taylor?'

She put the spoon down next to its brothers. 'Happy
news,' she said. 'He has married his cousin.'

Leonora stared. 'We heard nothing of an engagement.'

'It appears to have been a quick affair. How like Tom that
is – typically no-nonsense.'

'Mr Taylor, married,' Leonora said. After a moment she
asked, 'What is the young lady's name? A cousin you say?'

'Emma. Emma Taylor.'

Leonora's forehead creased, she shook her head, perplexed. 'It must be a branch of Taylors I'm unfamiliar with.'

'I have certainly not heard of her before this.'

'He didn't speak of her?'

'Never.'

'Most peculiar.' Leonora's eyes flashed. 'Deceitful, almost.'

'No,' Ellen said. 'Not that, surely.'

'Well.' Leonora's shoulders drooped. 'We must be glad for
him, I suppose. He is of an age to be married. A physician
should have a wife.'

'We must wish them every happiness.'

'Yes, yes … of course. Though one would always ask for
notice of so important an event. I suppose this is why we

haven't seen him in such a while. Will you write to him, Ellen, and congratulate him and Mrs Taylor on my behalf?'

'I'll do so this evening.'

'You'll express it better than I.' Her eyes widened. 'Perhaps he will bring her here.' Then, grimly, 'God save us from that.'

They looked at each other. 'I hope she's … I hope she's his measure. That she fosters his interests, understands him, as such a man deserves—' A sudden cough broke her voice. Leonora stood, put out her hand as though to steady her.

'My dear …'

A quick knock, and Kate bustled in. Ellen turned away. 'Joanna looks poorly, Miss, and complains of cramps. She's white in the gills. Have I leave to send her back to her bed?'

'I'll come and see her myself,' Ellen said. 'In a moment.'

'Very well, miss.' She heard the door close again, softly.

No more was said that day about Mr Taylor and his new bride. If Leonora mentioned anything to Jack, he had enough tact not to bring it up with Ellen. Though he seemed gentler over the next weeks, refrained from snapping or demanding things from her. She caught them exchanging glances, affecting cheerful voices whenever she came in the room. Their solicitude brought her closer to breaking than anything else might have.

In the first week of July a message arrived from Cork: could they send the carriage to collect Sam, home at last from Trinity College?

Ellen waited by the parlour window. The day was fair, with a breeze from the south. As soon as the carriage pulled up, a tall young stranger jumped out, dressed in a dark blue jacket and cream britches.

She forgot herself and ran outside, towards where he stood on the gravel. She called, 'Sam.'

'Ellen? Of course, sister, it's you.' He reached for her hand, smiling broadly. His attractiveness was all that of youth: clear grey-blue eyes in a smooth face, wayward hair that obviously defied his hairbrush, long limbs, loose with the quivering, unrestrained energy of a yearling.

'You have Jack's features,' she said. Though his gaze had a frank clarity that Jack's lacked.

'Emanuel always says I look like you,' Sam said. 'I see I am fortunate indeed.'

'Flatterer.'

He laughed. 'I want to see everything,' he said. 'I've been dreaming of the sea.'

'The countryside shows itself at its best for you,' Ellen said.

'Is it not like this all the time?'

'Come again in January,' she said.

'I well remember what winter here is like. What's this?' He drew his thumb across her cheek. 'No womanly tears, sister.' She couldn't speak, shook her head. 'I'm heartily glad to see you also,' he said.

He managed to sit still for an hour, for Leonora's sake. Then he leaped to his feet. 'Jack, you'll take me for a tour of the kingdom?'

'I no longer drive.'

'What? Why?' No reply. He shrugged. 'No matter. I shall drive us.'

Jack waved towards the door. 'Take Ellen.'

'Nonsense. I want you to show me the land. You're the master here.' Jack sat further back in his chair. Sam offered his arm. 'Come. Come!'

A pause, during which Ellen waited for the storm to break: *Leave me be, damn it.* Then Jack placed his hands on the armrests. 'Very well. Though you'll have to carry me to the gig. If you can.'

They disappeared for the rest of the afternoon. When they returned, dusty, sweating, Jack had been clearly won over, his smile a crooked reflection of his younger brother's. At dinner, Sam said, 'We'll have time to get acquainted this summer, sister. I expect you to show me the countryside.'

'I hope you won't be bored. There isn't much amusement here.'

'Amusement!' He threw down his napkin. 'Amusement is precisely what I need a holiday from. I've had enough of that in Dublin.'

'I guarantee you'll be craving it again come September, after a summer spent in the wilds of nowhere.'

'Is that not the point of a holiday? Respite from what we know, a change in routine?'

'I wouldn't know. My routine never changes.'

'Surely you have no need of change, living here in paradise.'

'Even paradise grows tedious in the end.'

He changed the subject. 'You're quite famous, I hear,' he said.

'Who says so?'

'Our cousin Tom Taylor. He asked about my "brilliant sister". I visited him in Dublin, as I thought I should. You know we met him when he came to London.'

'He exaggerates.' She took a sip from her water glass.

'You saw Mr Taylor?' Leonora leaned across the table. 'We heard he got married. Did you by any chance see the new Mrs Taylor?'

'Certainly,' Sam nodded. 'She was at home when I called.'

'Well, tell us, pray. What are her looks?'

'Unusual. Taylor says she was born in India, like himself. Dark eyes, dark hair. Small, delicate. I suppose some would think her pretty. A trifle doll-like for my tastes.'

'And her demeanour?'

'She didn't say much. My impression was that she was rather in thrall to her husband. But then, Taylor is a charismatic fellow, would you not say, Ellen?'

'I suppose so.'

Leonora cleared her throat. 'Would you say it's a good match?'

'I'm no expert,' Sam said. He took a long drink of wine; looking up at the avid faces, he relented, shrugged, 'Good enough, as far as I could tell. She was educated in Ireland; apparently they have much in common. And, I believe she has money.'

Silence, while this was digested. Then Ellen said, 'Money wouldn't matter to Tom.'

'It would matter to any man.'

After a moment she said, 'I wrote to offer our congratulations.'

'He mentioned that. He asked me to tell you that he would go to Kerry this summer to collect mosses, and to write him if you want samples.'

'Those mountains are full of wonders. I should like to go there myself.'

'Not while your headaches linger,' Leonora said.

'Headaches?' Sam said. 'Are you unwell, sister?'

'I work late into the night and tire myself, that's all. Illustrating plants is close work and the candlelight is inadequate.'

'She's blinded by pain, for days at a time,' Mrs Hutchins said. 'Lies on her bed with the curtains drawn. When she recovers, she stays up all night reading. It can't be healthy to be so consumed by books as your sister has lately become. Perhaps you can caution her to preserve her health. She doesn't listen to Jack or myself.'

'Dear mother,' Ellen said, exasperated.

Sam had stopped eating, his head turning to watch each as they spoke. He readied his fork again and dragged across a slice of mutton.

'What are you reading, sister?'

'Dante, at present.'

'Dante!' Sam said. 'How came you by Dante?'

'Mr Dawson Turner, a botanical acquaintance of mine, sent me a copy,' she said.

'What do you think of the *Inferno*?'

'Delightful, glorious, horrible by turns.' In fact, she was contemplating abandoning it. Its darkness was more than she could manage at the end of a long day.

'Does Mr Turner regularly recommend books to you?'

'Occasionally.' She poked at her peas, sending them rolling around the plate. 'My education is lacking, as you know. There's no one else here interested in books.'

'Well,' he said, with his mouth full. 'I intend on reading nothing this summer. In fact, if I so much as see a book, I'm liable to hurl it out the window. Even the divine Dante. So keep him in your room, sister Ellen. This is excellent mutton, by the way, mother. The food in my accommodation in Dublin is miserly.'

'You look thin. We'll do what we can to fatten you up before you go back.'

Later he bent to whisper to Ellen, 'I plan on corrupting you this summer, sister, by dragging you away from your plants and books. I don't like the sound of those headaches.'

They had taken the boat to the beach below Blue Hill. After walking the shingle for a time, they climbed the gentle, curved hill and lay down, the ground beneath their backs dry and warm. A faint crackle in her ear: a beetle making its way along a stalk. The stalk trembled, bending under the creature's weight. Its antennae groped the air, like a blind man's fingers. She flipped her gaze upward; the sudden shift from tiny to the infinite brought the sky rushing down. Or was it she who flew, towards the twisting clouds? Intoxicating freedom, such as she hadn't known since the summers of her schooldays, when she ran in the back gardens of Platanus.

Sam had corrupted her, as he'd promised: she closed the door of her study, laid her letters aside unanswered. She rushed through her duties around the house, her work in the garden. Though utterly uninterested in how his meals arrived to table, how stacks of clean linen appeared daily in his room, Sam was happy to lug buckets of water, dig early potatoes. He had a talent for making a sport of things. They drove the gig as far and as fast as daylight and the rambling boreens allowed – the dry summer had baked the narrow mud roads into something like a fit surface, though they had to wrap their faces against the dust clouds that marked their progress on the mountain side – and took the boat out to Whiddy and Garnish islands, beyond the bay, to discover new beaches and coves. In only his drawers and undershirt, Sam waded into the sea, yelling in protest at the water's fierce cold. Afterwards he ran to where she sat on the beach. His wet underclothes clung to his slim body. He shook his hair, sending a cascade of drops showering over her; she squealed obligingly, bid him mind his manners. He declared her as good a companion as he could hope for: 'sensible, for a woman, with spirit aplenty'.

He seemed younger than twenty-three. He talked of Trinity, friends he'd made, his plans after university – first, travel; like Captain Cook he had a desire to see the world ('you can accompany me, Ellen, and document the botany of the lands I discover, as Sir Joseph Banks did.') Emanuel, he said, expressed a desire to travel also.

'He discusses such things with you?'

'He's not so implacable as he pretends,' Sam said. 'I can usually get around him.'

'What of his disagreement with Arthur?' She hoped Sam's open nature might lead him to disclose more. Instead he made a show of blowing seawater from each nostril. Then he fell on the shingle, leaned back on his elbows, gazed out to sea. 'That will blow over,' he said. 'Besides, Arthur is a dull dog.'

'That isn't fair,' Ellen objected. 'Arthur is responsible, sober. And kind. You haven't spent enough time with him.'

Sam nodded, ever fair-minded: 'Possibly so. And what of Jack?'

'Jack has a good heart.'

'And a short temper.' There had been a couple of minor eruptions, laughed away by Sam, who refused to quarrel.

'His infirmity frustrates him,' Ellen said. 'Thoughts of what his life might have been. He'll never have your confidence. Years of associating with privilege have given you ...' She fumbled for the right words. 'A certain refinement.'

He hooted at this. 'Emanuel would disagree. He considers me about as refined as a turnip.'

'You're impetuous, selfish, as young men often are when the world is only agreeable. Yet you also have an ease, that education brings.'

'And what of you?'

'What of me?'

'Are you happy here, nurse to Jack and our mother?'

Water winked on her sleeve where it had landed from his hair. She scooped up a handful of pebbles and let them fall through her fingers. 'Happy enough,' she said.

'You don't ask for anything else? There's nothing you want?'

'No. Except ...' She stopped.

'Except?'

She avoided his eyes. 'I thought at one time I should like to have a child.'

A pause. She wondered if she'd embarrassed him. Then he said, 'You may still marry.'

She shook her head. 'I feel the chance has passed.' He didn't reply, seemingly absorbed in the view out to sea. 'You know,' she said, 'I am twenty-four. The same age now as our sister Katherine was, when she died.'

He shrugged. 'I don't remember her.' His tone was unconcerned.

'I suppose you were too young. I barely do myself. Does it not strike you though, as unimaginable? How in an instant …' Her voice trailed away.

'I never think of such things.'

'No?'

'No,' he said firmly. 'And neither should you. Today is enough to contend with. And today we are alive. I can feel the blood in my veins, the light on my face, my heart beating in my chest …'

'Speaking of time passing,' she said, keeping her voice light, 'half the summer's gone, and I've done nothing. With the weather perfect …'

'I'm a distraction.'

'It's my own fault. I lost energy for the task. There seems no point, somehow.'

'Any particular reason?' he asked.

'No.'

'Mother says that you've been out of sorts.'

'Oh?' She looked at him, surprised.

'She's concerned. That you may have been … disappointed, in some way.'

Ellen flushed. 'I don't know what she means.'

'She was so vague I hardly understood her myself.' He shrugged. 'You know Mother.'

'Yes.' She could well imagine Sam's mystification at Leonora's veiled interpretations of Ellen's state of mind.

He leaned his cheek on his knee, looked at her sideways. 'So naught's upset you?'

Casual dismissal, she thought, was more credible than out-right denial. 'Oh, a small thing.' She gave a little laugh. 'A trifle. Womanish nonsense. Offence where none was intended. I've forgotten it.'

'You won't say more?'

'No. Don't ask, I pray.'

'But I may tell Mother if she asks that all is well?' He looked as eager as Lettie, when she ran ahead and stopped to look back at Ellen's slow progress: *Why do you dawdle? Will you not keep up?* Absorbed in himself as he was, Sam would not wound her by clumsily stabbing for the truth. It would pass from his thoughts as quickly as pipe smoke. 'All is well,' she said.

'Very good.' At that moment, more than when he'd asked before, she felt like clutching his arm and giving in to a storm of child-like weeping.

A group of men came into view, dragging a boat down the beach. Sam watched them for a time, shading his eyes with his hand. He jumped to his feet, pulled on his coat. 'Come. They will say the Hutchinses are an indolent lot; we will give cause for resentment.' As they walked, he stared after the fishermen as though he longed to run and join them.

Before dinner, as they played cards in the parlour, Lettie began frantically yipping. Sam went to the window.

'A messenger, I think,' he said.

Knocking on the front door, footsteps, muffled voices. A hollow slam, more footsteps. The parlour door creaked open and Kate came in.

'For you, ma'am,' she said, handing a letter to Leonora.

'We aren't expecting anyone, are we?' Leonora eyed the letter as if it were a small rodent escaped into the room, unwelcome and potentially irksome. She passed it to Sam. 'Open it, tell us what it contains.'

He read aloud.

Waterford

Dear Madam,

I believe our friend, Mr Dawson Turner of Yarmouth, advised Miss Hutchins that our group might call on you at Ballylickey House on our way to Kerry. At present we are in Waterford; we plan on setting out early tomorrow morning to arrive in Cork sometime next week.

Our intention is to make the acquaintance of Miss Hutchins, who has been so helpful to us in our botanical studies, and see some of the sites of interest locally before we travel onwards. I trust, madam, that we will not put you to any inconvenience, and urge you not to take any special pains. Mr Turner has written a letter of introduction on our behalf, which is enclosed.

I will send a note once we've reached the city of Cork to give notice of our arrival in Bantry.

Your obliging servant
Lewis Dillwyn

'Who is Lewis Dillwyn?' he asked.

'A botanist correspondent of mine,' Ellen said. 'A friend of Dawson Turner. He mentioned they might call on us.'

'What in all heaven will we feed them?' Leonora said. 'In this weather we can't keep meat in the house. Is there enough wine? Sam and Jack between them have surely drained our reserves.'

Sam had unfolded a second sheet of paper and now held it up. 'Here is Mr Turner's note. *I ask that Mrs Hutchins receive my friends Mr Dillwyn, Mr Leach and Mr Woods. We are of long-standing acquaintance and I attest to their characters unreservedly, each being an educated gentleman of good family et cetera, et cetera …*'

Ellen pulled the letter from Sam's grasp, reading through it while he whistled, 'What esteem they must hold Ellen in, to take time to come here. An honour indeed.'

'A nuisance, you mean,' Jack said. 'More twittering Englishmen, come to gawk at the Irish peasant and condescend to the landowner.'

'Such hostility, brother. We must extend them every courtesy for Ellen's sake.'

'Naturally,' said Leonora. 'But what, I ask again, of meat? They will certainly stay for dinner.'

'We can kill a couple of chickens,' Ellen said. 'There's ample vegetables, greens, fruit. We'll have cream in the morning. As for wine – even Sam has not managed to completely deplete our stock. It will suffice.'

As if she had plunged her head into cold water, her torpor lifted. She left her mother and brothers and went directly upstairs.

The air inside her study smelled, not unpleasantly, of rotting leaves and damp soil. She pulled up the sash a couple of inches only, in case her papers might be disturbed. Each dried sample on its paper mount had the date of collecting marked neatly underneath. On the shelves of an old dresser she'd categorised her extra samples: *algae, lichens, mosses*. Where she'd found them, what she'd learned of each. She kept her drawings in a deep drawer. They now numbered in the hundreds. Taking out the top drawing, she studied it for a moment. Competent, certainly. Though looking at it now, it did not wholly reflect the satisfaction she'd experienced at the time of its creation. Still, she'd accomplished what she'd intended – the effect of seaweed underwater, floating, full of light, like a woman's hair in the wind. She put the drawing back, took one more look around the room, and closed the door behind her.

Kate would kill the unfortunate fowl when the time came, though Ellen thought she could do it herself, if needs be. You could do what you never thought you could, if you closed your eyes and did it swiftly.

NINETEEN

Mr Dillwyn proclaimed the breakfast at Ballylickey the finest he'd had since landing in Ireland.

'The food has otherwise been abominable,' he said, mopping at his forehead with a large handkerchief.

'Are you too warm, Mr Dillwyn?'

'Thank you, Mrs Hutchins, I'm quite comfortable. Your dining room is delightfully cool.' His coat was cut from a dark-red, suede-like fabric, and accentuated his high colour. Dark, curling hair circled a bald spot on his crown.

The detritus of breakfast – crockery, crusts, crumbs – lay scattered across the tablecloth. Sam had gone to Bantry early, on horseback, and met the Englishmen there; he'd accompanied their hired carriage back to Ballylickey where they consumed a large dish of scrambled eggs, bread, potato cake, a full slab of butter and half a jar of gooseberry preserve.

'We've had one or two decent meals,' Mr Woods said. 'Though the claret is often thinner than I would prefer.' He pulled on his nose and shook his head. Of a similar age to Mr Dillwyn – somewhere around the mid-thirties – his large body curled into itself as though to diminish its looming presence.

'What say you, Mr Leach?' said Jack, from the head of the table. 'How do you like the food in Ireland?'

'Oh, Mr Leach is not a gourmand,' Mr Dillwyn said. 'He'd scarcely notice if he ate offal or filet.'

Said affectionately: Mr Leach smiled. 'I think the food perfectly adequate.' His smooth face looked young enough to belong to a man still in the teens. He was thin and pale, with a distracted air. His light brown hair flopped over his forehead;

each time he spoke he brushed it back and held it off his forehead, as if to give air to his thoughts.

'Are you all botanists, like my daughter?' asked Leonora.

'I collect plants, mother,' Ellen said. 'I am no botanist.'

'Too modest, Miss Hutchins,' Mr Woods said.

'By any methods used to define a botanist, we can say that Miss Hutchins is one,' said Mr Dillwyn. 'In answer to your question, madam,' – he bowed towards Leonora – 'Mr Woods and myself are avid botanists. But young Leach inclines more towards entomology … would you believe a man could journey four hundred miles to find a particular beetle?'

Everyone laughed. 'It's true,' Mr Leach said, nodding around the table.

'And did you find it?' asked Leonora, blue eyes wide.

'I did, madam. *Sitaris muralis*. On a buttercup leaf in the New Forest.'

'And what did you do with it then?'

'I brought it back home in a tin box and added it in my collection.'

'Alive?' She sounded incredulous.

'Alas, madam, it was necessary to then kill it, before mounting it with a pin through the elytra.'

'How horrible,' she said.

'Indeed,' he said. 'But I did a nice drawing of it afterwards.'

'Well. I call it a strange pastime.'

'I'm utterly consumed by it,' he said. 'My medical studies seem dull in comparison.'

'And what is your profession, Mr Woods?' asked Sam.

'Architecture, sir.'

'An exciting field,' Sam said.

'Though similar to my young friend, my true interest lays elsewhere – hence this trip.'

'Botany certainly seems to exert a peculiar fascination,' Sam said.

'You don't share your sister's passion for plants?' Mr Leach asked.

'One botanist in the family is enough. Besides, I would never have the patience.'

'You may in time, sir. When youthful impetuosity passes away, giving rise to more reflective pursuits.'

'Perhaps you'll develop an interest in trees, like Arthur,' Leonora said.

'Have you chosen a profession, sir?' Mr Dillwyn said.

'Not as yet,' Sam said. 'At present I'm in Trinity College, Dublin.'

Mr Dillwyn bounced his fist lightly on the table. 'Every man needs an occupation. Not only that – a man should fill his spare time as fully as he is able. Even a lady, once her familial responsibilities are met, can make a useful contribution. And now ...' He pushed back his chair. 'We must get on. Or the morning will be entirely spent.'

'What are your plans, gentlemen?' Jack asked.

When he'd emerged from his bedroom that morning he'd shaved and put on a clean shirt and jacket; he even succumbed without protest to Ellen's straightening his cravat.

'We intend exploring the local beaches,' Mr Dillwyn said. 'In search of the fine sea plant specimens Miss Hutchins finds so regularly.'

'Perhaps we can visit some of the islands close by,' said Mr Woods.

'Whiddy, at least,' Ellen said.

'May I offer you the use of our boat?' Jack said. 'I have a man who would take you out at your convenience.'

Mr Dillwyn bowed. 'Most considerate of you, Mr Hutchins. We'd be delighted.' He drew back his chair. The other two men took the cue and with a last wipe of their mouths, stood to go.

'You'll come back for dinner,' Jack said.

'If your breakfast is anything to go by, sir, we'll not be kept away.'

'I'd enjoy a trip to Whiddy myself,' Sam said.

'The more the merrier,' Mr Dillwyn said. 'Miss Hutchins, you'll join us, of course.' Before she could speak, Sam said, 'Alas, there's only room for four passengers.'

Again Mr Dillwyn's hands flew apart. 'A pity! Still, Miss Hutchins, before we depart we'll hold you to your promise to show us your herbarium. A fleeting look only, as we've dawdled so long over breakfast.'

They proclaimed themselves astonished at the extent of her collections, crowding around her desk and talking all at once. It was midday before they made to leave. As they bustled out the door, Sam took her elbow.

'You don't mind my going? I thought it hospitable to offer to accompany them.'

'You were correct,' she said. 'Someone should.'

'And I'm not sure it would be seemly for you to go with them alone.'

'Kind of you to consider my reputation.'

He looked crestfallen, like a small boy. 'I meant well, sister.'

She sighed. 'I know it. Just go. They're waiting.' He looked at her again, uncertain. The Englishmen could be heard outside, chattering loudly. His face cleared, he nodded, and ran to join them.

Once they'd gone, the house rang with silence. Jack disappeared into his room, banging the door shut. Ellen paced the rooms, bothered Kate in the kitchen. 'Go for a walk, Miss, leave us to our work.' She fled with her book, found her way to the waterfall. But the words, magical as they were, proved too dense to sustain her interest; she threw *Don Quixote* down, peeled off her gloves, chewed on her fingers.

She knew every stone of Whiddy, every rock pool. Yet Sam saw it as his right to jump from his seat and seize the opportunity to escape with the Englishmen.

She couldn't, shouldn't blame him.

A sudden sharp sting and the taste of blood: she'd bitten through the nail. She wiped her fingers on her skirt, retrieved her book from the grass and went back to the house.

The kitchen windows were fogged from the heat of the fire. Joanna held the fleshy pillow of a dead hen, three-quarters plucked, under her arm. A stray feather had lodged in her hair. Half moons of sweat spread under her armpits. Kate sat polishing glasses; she set one down, spit on her cloth and picked up the next. Ellen examined a bowl of freshly picked lettuce. 'Wash these thoroughly, Joanna. No snails in the salad.' She lifted the cloth covering the cucumber soup, dipped in a knuckle and frowned. 'There's too much salt. Isn't there? Did you taste it?'

Kate swivelled in her chair. 'The same as always. Now don't work yourself into a state. All will go off beautifully.'

Mr Dillwyn pulled off his broad-brimmed hat and waved it at Ellen. 'See my red-faced companions, Miss Hutchins,' he said. 'A healthy outdoor complexion can be taken too far. Eventually the danger arises of looking like a farm hand.'

Mr Woods and Mr Leach followed him inside, blinking in the dark of the hall. As Mr Dillwyn said, their faces had darkened from sun and wind. 'A splendid outing, Miss Hutchins,' Mr Leach said. 'I think my friends are exhausted.'

'You have the advantage of youth, Leach,' Mr Woods said, rubbing at his eyes.

'Please refresh yourselves upstairs in my brother's bedroom,' Ellen said. 'Then join us in the dining room.'

In the kitchen, Kate stood over the golden, spitting chicken. Her blouse was soaked with sweat. The gravy was ready, already in its dish and close to the grate to keep warm. Joanna lifted the pot of steaming potatoes from its hook.

'We must serve soon, Miss, or everything will be burned, overboiled, or start going cold.'

'I'll get them to table now. Joanna, change your blouse – borrow one of Kate's – you must serve the wine, I can't ask Master Sam to do it in front of the Englishmen.'

'When we first arrived in Bantry we heard a remarkable thing,' Mr Dillwyn said. 'A howling in the street, like that of a pack of sorrowful dogs. Mourners at a funeral, just that minute passing through the town.'

'In all likelihood many of the chief howlers had no knowledge of the dead person,' Jack said. 'In fact, some are paid by the family of the deceased.'

'Extraordinary,' said Mr Woods.

'The common people here treat death differently,' Ellen said. 'It's shared by the entire community, not for the family to bear alone. There's an initial outpouring of extraordinary emotion. They lay out the body at home and watch over it all night, singing and telling stories.'

'And drinking whiskey,' Leonora said. 'Men *and* women. The Irish wake is no more than an excuse for drunkenness.'

'Maybe so,' Ellen said. 'But perhaps something is exorcised in the process.'

'What do you mean, Miss Hutchins?' Mr Dillwyn asked.

'Afterwards they seem to find it easier to move on, recovering from the most lamentable sorrow. They do not hold with long periods of public mourning, as we might.'

'Grief is a private thing,' Leonora said, wiping her lips. Her eyes grew misty. 'I can't see how keening, as they call it, for the whole world to hear, could alleviate it.'

'Well, one thing is certain,' Jack said, 'The Irish are obsessed with death. You won't hear better ghost stories anywhere else, gentlemen. Spirits haunt every crossroads. I pray you don't meet an old crone or cloaked stranger on the road back to Bantry later tonight.'

'The membrane between this world and the next seems thinner in this part of the country,' Ellen said. 'Structures, features of the landscape, have an impermanence about them, as though they could dissolve.'

'Come, Miss Hutchins. Surely you, a lady of science, don't give credence to spirits?' Mr Dillwyn said.

'She has imagination,' Leonora said. 'Too much of it.'

'What think you of our landscape?' Sam asked. 'In comparison to England?' Ellen shot him a glance of gratitude.

The gentleman hastily swallowed a mouthful of roast chicken. 'Lacking, Mr Hutchins,' he said. 'Decidedly lacking. There are no trees, no great woodlands! And the mountains barely register as such. Little more than hills.' He wagged his fork. 'Another thing that we have been struck by: the looks of some of the women.'

Mr Woods stiffened. Mr Leach's fork suspended mid-air on its way to his open mouth.

'Explain, Mr Dillwyn,' Leonora said.

'Well, madam, it's an extraordinary thing.' He nodded vaguely in Ellen's direction. 'The handsomeness of some of the ladies we've met on our travels is almost directly, inversely, comparable to the ugliness of others. The looks of the lower classes in particular verge on caricature at times. Crudely assembled limbs, bulbous eyes, grotesque deformities ... scarecrows and hags abound.' He paused for air, and a gulp of claret.

There was general silence – amused, appalled – for a moment, then, 'You surely thought the countryside around Bantry picturesque, at least?' Jack said.

'Exceptionally so,' Mr Woods said quickly. 'The bay is glorious.'

'But there again,' burst in Mr Dillwyn. 'Where there is some beauty of landscape, there's a lack of attractive, well-designed settlements. There are few fine houses such as Ballylickey.' He waved his knife, taking in the dining room with its dusty landscapes and sturdy old furniture. 'Some of the great seats we've visited, like Bantry House, for example, are the equals of any in England, to be sure. But where are the dwellings of the middle classes? The merchants, the better tradesmen, the prosperous farmers? They don't seem to exist. And the dwellings of the poor! I've never seen such poverty and squalour.'

'It's true,' ventured Mr Woods, 'there doesn't appear to be a middle class in Ireland. People are either exceedingly comfortable or impoverished.'

'There's great poverty here,' Sam said. 'I'm struck by it afresh every time I return. But that can be blamed in large part for the unjust way the native Catholics are treated. The worst of the Penal Laws linger on in practice, and are burned into the memories of the people. The system of land ownership makes it all but impossible for them to prosper.'

'Something my husband tried to counter against in his lifetime,' said Leonora.

'Granted there's something in what you say,' Mr Dillwyn conceded, in the generous tone used by a man confident he has the superior argument. 'But there can be little excuse for the general filth and *laissez-faire* attitude we've consistently encountered in every inn, tavern and shop. Not to mention the blatant opportunism and over-charging, hand in hand with a kind of sly obsequiousness – surely this is indicative of an endemic national characteristic?'

Kate came in and began clearing away the second course. Mr Dillwyn leaned away to allow her take his plate.

Jack cleared his throat. 'I don't recognise your description of the Irish as applying to anyone who works for this household,' he said.

Sam spoke up. 'I admit that there can at times be a more casual approach to service, Mr Dillwyn, in comparison to what you might be used to in England ...'

'Casual is not the word, sir!'

'... but again, some of that can be attributed to the Catholic Irishman's sense of himself as a second-class citizen in his own country. It has bred a degree of resentment and hostility, for which the English parliament most take some responsibility.'

'I'm sure what you say is true, sir,' Mr Woods said, looking pointedly at Mr Dillwyn. 'My friend is simply speculating as to the reasons for some of the behaviour we've encountered. Naturally, the visitor is at risk of jumping to conclusions that are perhaps too hastily judged.'

'I still say there must be an intrinsic weakness of character that allows a people to accept such low standards. By doing so, the Irishman does not promote his cause.' Mr Dillwyn shook his head.

'Do you include us in your condemnation, Mr Dillwyn?' Ellen said.

He blinked at her, as though he'd forgotten she was in the room. 'Miss Hutchins?'

'Our family has worked side by side with the people whose values you scorn, endeavoured to treat them fairly. We cannot agree with your caricature of our servants and neighbours. You forget that we ourselves are immersed in this place. It's not simply a remote corner of England, to be judged as such.'

He frowned, digesting this. 'I think you misunderstand me, Miss Hutchins. I don't accuse you, or anyone of this house, with a lack of industry or attentiveness! If all followed your example, I would have no cause for comment.' He spread out his hands in a conciliatory gesture. 'And comment is all I offer. No doubt it seems harsh. Often those living in the middle of things cannot see what must be obvious to anyone coming from the outside.' Mr Woods threw down his napkin and rolled his eyes. 'Forgive my frank manner,' Mr Dillwyn went on, unconcerned. 'I'm told it can be somewhat brutal. My wife – that most modest of women – cautions me against my honesty, which I find difficult to suppress.'

'Well, I for one, find Ireland delightful,' Mr Leach burst out.

'As do we all, dear Leach,' said Mr Dillwyn. 'My word, have I given the opposite impression?'

At that moment Kate and Joanna entered the room carrying a cherry tart – 'From our own trees, gentlemen,' Leonora said – with jugs of cream; Mr Dillwyn fell silent, following the women with his eyes as they set the platter down in the centre of the table. Mr Woods changed the subject to architecture; he asked for quill, ink and paper to be brought to the table and quickly sketched his uncle Jonathan's house, Paradise Hall, in whose design he'd had a hand, 'at the tender age of seventeen'. His voice rose higher with emotion.

'A fine villa, of every conceivable comfort and amenity,' he said. 'Alas, it is now lost; my uncle having financial difficulties, had to sell it.' He stared at his drawing morosely and covered it with a saucer.

The front door lay open to the mild night air. The waters of the bay showed as a distant sliver of glistening blue-black. Shapes within the garden were etched in grey light. 'A full moon, gentlemen,' Sam said, peering into the sky. 'You should have no difficulties getting back to your inn.'

'Be sure to breakfast here in the morning,' Ellen urged, 'Before you set out for Kerry. You pass by this way anyway.'

'You are kindness itself,' said Mr Dillwyn. A sheen of moisture covered his forehead and upper lip. The red of his cheeks had spread, mottling his entire face. He covered his mouth with his hand, his face contorting into a yawn. 'Ah! Excuse me, ladies. Goodnight.' He pressed Ellen's hand and tottered onwards towards the waiting carriage.

'Goodnight, goodnight.' Mr Leach, bowing, beaming, as he passed.

'Goodnight, Mr Leach.'

Lastly, Mr Woods, stooping, compacting his neck, as though in fear of hitting his head against the lintel. 'I wondered if you might like this?' he said, holding out his sketch of his uncle's folly. 'As a souvenir of our visit.'

Ellen took the sketch. 'Thank you, sir.'

'It's not accurate,' he said. 'Or particularly skilful. At my desk, with enough time, I'm draughtsman enough to produce satisfactory work. Though I'm not as proficient as you are, Miss Hutchins. Mr Turner was in raptures over your illustrations when I spoke with him last.'

'I'm improving, I think,' she said. 'I seem to have a better feeling for it than previously.'

'Constant application is the way forward,' he said. 'As with most things.'

'Woods,' Mr Dillwyn called from the carriage. 'Say goodnight, like a good fellow. You delay our hostess unduly, besides keeping your companions from their sleep.'

'In a moment, Dillwyn,' he said. He leaned closer to Ellen, said quickly, 'Mr Turner's regret that he could not come was palpable. But there's upset in his family at present.'

'I know of his troubles.' She bit her lip. 'Of course, I would have been glad to meet him. Perhaps in the future ...' Her voice trailed away. She put her hand on his arm. 'Mr Woods. Tell me – what does Mr Turner look like?'

He stretched his eyes in surprise. They were red, bleary, from the claret, or the effects of the day spent on the water; even hardy men – sailors, fishermen – could succumb to sun-fever. 'Well, I suppose most would say he is a fine-looking man. His countenance reflects his personality: amiable, intelligent. The best, most generous of men, and the kindest of friends.'

'That is my impression,' Ellen said.

'On my return to England I shall be able to tell him ...'

Again, Mr Dillwyn called. Mr Woods feigned not to hear. '… I shall tell him his impressions of you are quite correct. If anything, he underestimates your virtues.' He drew himself to his full height, bowed, and walked quickly out to the carriage.

❧

She had laid aside specimens for them to take away, offered them anything they might want for their own collections. Would Mr Dillwyn oblige her by sharing with their mutual friend in Yarmouth? Mr Dillwyn nodded, yes, certainly he would. He licked his fleshy lips and jostled against Mr Woods as they clustered around the table, dividing the spoils. 'Too much, too much, dear Miss Hutchins,' he said, packing the specimens into a small wooden trunk.

'You have more use for it than I do, sir. Besides, I can replenish most of it.' In truth, the rarer plants would not be found again so easily – *Hyoscyamus niger*, from Whiddy Island, *Ulva fistulosa*, from the beach at Ardnaturris, near Ardnagashel … What of it? Mr Dillwyn was welcome to whatever caught his eye, with all her heart. And if he happened to mention the source of the specimens to his friends – well, it was one way to ensure her name might circulate at the highest levels of botanical study.

Vowing to see the Englishmen safely over Priest's Leap and onwards to Kenmare, Sam had hired a team of fifteen local men to push the carriage up the steep mountain road. He reported back how, once they'd reached the summit, the three naturalists had jumped from the carriage and set out to explore the mountain. Here Sam had left them, scampering over rock and shouting to one another in delight about the abundance of *Saxifraga hirsuta*.

TWENTY

Mr Turner's eldest daughter loved shells.

There were many beautiful specimens on the beaches around Bantry. Mr Dillwyn had thought them pleasing, at least. His views on Ireland were harsh; it pained the Hutchinses to hear them, as it would a doting mother told ill of her favourite son, however she might secretly admit his flaws to herself. Yet despite his lack of tact, Ellen could not help liking him. He had energy and a kind of ruthless vigour she found appealing. She could forgive a person any number of sins once they were amusing or interesting: rather a stimulating transgressor than a dull saint. She walked the middle ground, anxious not to step too far either side, striving to be a good sister, a good daughter. Though her thoughts rebelled, no matter how she tried to rein them in. Thoughts of running away, casting off the cords that bound her.

Escape.

A shell was easy to collect, easy to keep. Dry as bone, small, light. When she'd gathered enough, they clinked in her pockets, making music as she walked.

She went out alone, walking, telling Sam he was too much of a distraction. For a time, the surge of blood – the sap of life – through her veins drove away the memory of pain. Then, the first twinkle of headache, like a distant, sickly star.

It heralded the raptor, cruel and merciless, that began its descent from the grey sky.

First a black speck, high in the distance. For a time, hope: that it might go again, disappear back whence it came. It grew larger, hovered. A clamp on her skull was attached by fine wire to its claws, that pulled, pulled, until she could not see, to look into her microscope, or take up her brush. And so she walked. Movement helped and the rush of the wind obscured the hammering in her brain. And if she stumbled, fell to the ground – there was no one to witness.

TWENTY-ONE

Sam left for Trinity at the start of September, with a parcel of best-quality linen shirts (seventeen pound's worth – a large dent in their accounts), Kate's seed cake, a case of wine and a five-pound note. His skin had freckled, his hair leached to pale gold. He waved wildly, then beat the side of the carriage through the window: time to go. Already he looked forward, to the road ahead, to the college term.

A cloud of dust obscured the carriage as it rattled down the drive. Leonora pressed a handkerchief to her eyes, then tucked it away, lifted her chin and went into the house. Ellen stood for a while, searching the sky for the white disc of light behind the morning mist.

He took the best of the Hutchinses with him. All they had been, and could be.

There were whales in the bay.

Dan said they would move on in the night; melt away as swiftly as they'd arrived. Yet they stayed into the early autumn, blue humps like new islands from as close to the coast as Bantry harbour. Static giants, churning the water with the rise and fall of their great tails. The fishermen were wary and navigated well away.

No rain for weeks. Ellen hoisted bucket after bucket-load of water across the yard to the garden. The chickens bathed in the orange dust, their strutting subdued. The sun slipped lower, changing the light, gilding the edges of bright objects. Silver mugs on the dresser, her locket. Joanna and Kate escaped the swelter of the attic and slept in the kitchen. At night the windows were left open, despite the fear that robbers might scale the front of the house, murder them all in their beds.

For a time, her headaches eased, restricted to periods of fatigue.

Her desire to collect came back. She'd exhausted the seaweed varieties and decided to increase her focus on lichens and mosses, inspired by Mr Turner's own efforts. These curious plants smothered branches and marked stone like ancient runes. A patch of bright yellow on a gable, illuminated by the evening sun. Flat rock patched with greenish white. *Verrucaria punctiformis* on an ash tree in the garden; *Opegrapha calcarea* in the woods close to the house; *Orthotrichum diaphanum* on a tree in Bantry churchyard. Less satisfying to draw than seaweed, though: her efforts resulted in murky, formless blotches that resembled mould on the page.

She consulted her volume of Withering. He advised as to their preservation: *The Musci (Mosses), which constitute the second Order of the Cryptogamia class, being very numerous, and mostly very minute, may be kept in papers folded to the octavo size. It is sufficient to place them in the papers, and to give them a moderate pressure for a short time. They dry readily and are not apt to spoil ... The Lichens require no care in drying; they should not even be pressed, or put into papers, but placed in shallow closed drawers, which are divided into small partitions ...*

His descriptions were curiously lacking, as though he too were somewhat perplexed by these most mysterious of plants. Beyond the pleasures of colour and form she came to be fascinated by their nature, how they attached themselves to a host plant yet did not take from it. Tenacious, complex: survivors, above all. Disregarded. To the untrained eye, barely qualifying as living organisms.

Overcoming their dread, the fishermen succeeded in harpooning one of the whales. The creature may have been sick, or dying, for it made little effort to fight. It was floated to shore, with a forest of pikes in its back, behind a legion of boats. Twenty men hauled it onto the beach, drag marks in dark blood staining the rose-coloured sand. For days they worked, slicing away the blubber, boiling it down in whatever pots they could muster, over regulated fires, tempered to allow the fat render into oil. Slabs of skinned blubber ranged along the beach, stacked in piles, ready to be added to the pots. A vile smell – a brew of fish, lard and cooking flesh – shimmered over the town.

Ellen found the scene horrible, almost biblical, like a depiction of the day of judgement. Yet it drew her in. Her handkerchief before her face, she ventured as close as she could without becoming overcome.

'Would you like a piece of skin, Miss?' A young woman, streaked with soot, red-skinned as though burnt, held up a knife. 'A keepsake.'

Large, straight teeth shone white in her dark face. Ellen nodded. The woman seized a piece of blubber with one hand and began sawing away a length of skin with the other. Her rolled up sleeves revealed the taut muscles of her arms. 'There you are, Miss,' she said. She could have been a merchant displaying a bolt of rare fabric. The length of grey-blue skin had the dead sheen of marble. Ellen prodded it with a bare finger. A shudder ran through her; for a moment she remembered the caul, its alien texture and repellent origin. Yet when she went further, rubbed the skin with the back of her hand, it felt smooth, as a woman's cheek. An archaic flag, she thought, of some primitive country. She had a sudden image of the great creature whose skin this had been, plunging through depths to where light vanished and the dark held sway.

One man circled the fires, adding fuel or dousing down where needed. Columns of smoke rose into the sky; the air seemed liquid. Ellen made an effort to smile as she took the skin. 'Thank you.'

When she showed it in the kitchen at Ballylickey, Joanna shuddered. 'Nasty thing. Phew, how it smells.'

'The butchery on the beach was horrible,' Ellen said. 'I felt ...' She struggled for the word. 'Shame.'

'That creature is a boon to the people around here,' Kate said. 'Come spring, they'll live off that oil. What's one dead creature in a sea of creatures? And how is it any different from gutting a pig, or a salmon?'

Kate was right, she knew. The people had no guarantee of food in their bellies, and had a right to harvest whatever they could.

The remaining whales left within a week. Perhaps the slicks of blood that still floated in the bay had caused them a kind of primal distress. Or, as Dan said, it was simply time for them to go.

∽

The beauty of a slab of rock on a December morning, covered in a variety of mosses and caught by a beam of low sun, stabbed her with a clarity akin to pain. She stood, transfixed by the shades of evergreen.

It had become increasingly apparent that the Withering microscope given her by Dr Stokes was no longer adequate to the task. The leather box had frayed from constant opening and closing. A screw had come loose and rolled away, forever lost between a gap in the floorboards.

She spoke to Jack. Could the cost of a new microscope be justified? He said nothing for a moment, squinted at her in the manner he had when thinking about money.

'The one I have is limiting my success.'

'Hmm.'

'I realise there's more urgent need elsewhere.'

'Always.'

'But—'

'Yes?'

'I need it.' Silence. 'I believe we can find the money for it. I have asked for very little since I came here.'

'Have you something in mind?'

'I asked Mr Dillwyn while he was here to recommend a superior model. He was shocked at how much I'd achieved with the basic one I have.' She hesitated. 'He didn't mention cost.'

'No doubt he thought it irrelevant. As it would be to that gentleman of means.' He frowned, rubbed his nose. 'Very well.'

'I'll leave you to consider it.'

'I've decided.' He smiled crookedly at her expression. 'It's true, you never ask for anything, not dresses, bonnets, or jewels. If it doesn't require us selling off the furniture, I say you shall have it.'

It arrived a month later, securely packaged within a crate of straw. An ornate label announced the name of the retailer: *R. B. Bates, supplier of Mathematical, Optical and Philosophical Instruments, 17 Poultry Lane, London.* Jack had found the money somewhere; there had been no need to sell anything, though at fifteen pounds it cost as much as a horse.

And so much gleaming brass and glass: at first, she was almost afraid of it. She wiped continuously at its perfect surface, reluctant to leave even a thumbprint. Then she scolded herself. It was a tool, to be used. Hadn't she earned the right to dominate it, the knowledge to manipulate its screws and bolts, its gliding parts?

The model name – the *Jones Improved* – had a matter-of-fact ring she approved of: determined, utilitarian. A handsome wooden base lodged the pillar. On top of the pillar, the brass tube through which the transformation worked itself. Adjusting the stage, examining a specimen under the glass for the first time, she drew a sharp breath. The image was vastly superior to that she was used to, every detail clearer and sharper, magnification glorified beyond her imagining. Her mind raced with the possibilities of spring. She had been given new eyes, new powers.

TWENTY-TWO

'I have decided to go to Dublin.'

She had been holding back the words for days; now, at the breakfast table, they burst from her. Jack splashed his coffee.

'Whatever for?'

'Dr Stokes has written. A Swedish botanist, a Mr Agardh, is visiting Ireland. He wants to meet me. Besides, I'd like to see the Stokeses again.'

'But the journey ...' her mother began.

'I haven't had a headache in weeks,' she said. 'I feel I am able for it. Mother, you are well enough. I know it will be dull for you, Jack, but I'll stay only a short time.' Neither spoke. Jack crumbled his bread to pieces, stared at his plate. She said, gently, 'I'm going. Before you know it, I will return.'

'Well. I expect the change will do you good,' Leonora said. She put her hand on Jack's sleeve. 'Don't you think so, Jack?'

Ellen walked around the table and kissed the cotton cap covering her mother's hair. 'Thank you, Mother.'

'You'll come back to us though, won't you, my dear? No matter how much they ask you to stay?' She grabbed Ellen's hand. Jack's eyes asked the same.

'There's no question.'

Still, the evening before she left, he sulked in his room until dinner. When he emerged, he took his place at table without speaking. Kate brought in the soup and set it down with a bang so that it slopped over the edge of the tureen. She too thought the trip ill-advised and had said so, repeat-edly. Miss Ellen would have to mix with common sorts on the road; there was nothing to eat in the inns except green potatoes and dry bread. If Miss Ellen survived the thieves and rogues to get there, Dublin itself was a pit of pestilence and disease. She would be overcharged by merchants and exposed to God-knew-what kind of low manners. Kate had never been to Dublin and never would; two hundred miles away or two thousand, she'd not soil her boots with the filth of its streets.

Ellen mopped up the spilled soup with her napkin lest it stain the tablecloth.

Arthur came from Ardnagashel to take her to Cork, from where she would get the day coach.

'I'll try and call in more frequently,' he said, as the carriage rolled along the main road.

Ellen looked out at the countryside. A ragged family walking the road – man, wife, two thin children – stopped and stood in by the hedgerow as they passed.

'You're quiet. You're not regretting your decision to go?' Arthur asked.

'My regret will be only the deeper if I don't,' she said. He nodded, seeming to understand.

Once in Cork he went to the offices of the Royal Mail Coach Company and purchased her fare. He had arranged accommodation overnight in an inn, a clean establishment with light-filled rooms and plenty of hot water, beside the River Lee. At half-past six the following morning he deposited her at the coach station on Patrick Street. His eyes narrowed as he scrutinised the other passengers, a respectable-looking assemblage heavily wrapped in cloaks and shawls in provision for the worst of weather. Seemingly satisfied, he offered all a curt 'good morning' and helped Ellen onto the seat.

'You'll write once you get to Dr Stokes's house,' he said.

'I promise,' she said.

'Then goodbye, sister.' He took from his shoulders the wool wrap he habitually wore, ever afraid of chills and agues. 'Take this. The wind from the mountains is pernicious and the coach windows are flimsy.'

He pushed it at her, walked away. Ellen had not expected him to wait until the coach departed, but still smiled to herself at the brusqueness of his leaving. So like Arthur – blunt, avoiding anything approaching sentiment. But unable to restrain his inclination to kindness.

They stopped in Kilkenny to dine and sleep. The inn had low ceilings and small windows, the gloom alleviated somewhat by cheap-smelling candles. Downstairs consisted of one large, untidy room. The patrons – mostly men – shouted across long wooden tables, swigging tankards of ale and wine. Peat smoke from the open fire made her eyes water. Scrawny hounds sloped and skulked; she saw one snatch a ham bone from a plate. The ladies present kept to the corners, some nursing small children. Older children ran about, cuffed aside by elbows and snouts. Having drunk a small glass of ale and eaten a plate of stew (as unpalatable as Kate had threatened), she escaped upstairs to her room. Minutes after she lay down on the straw mattress, her ankle began itching, then her arms. She searched about the bedclothes, found a scattering of red-bodied insects, like minute clots of blood; disgusted, she flapped them away. She imagined they crawled back as soon as she closed her eyes. Finally she wrapped herself in Arthur's blanket, though she still twitched without rest until dawn.

Back on the road the following morning, exhausted, pinioned between a large-bosomed woman and the window, she fell into a broken doze, jolting awake whenever the coach stopped: Royal Oak, Leighlinbridge, Carlow, Castledermot, Kilcullen, Naas. Passengers climbed off and on, treading on her feet, jostling her with their boxes and packages. On the last stretch she slept uninterrupted, until woken by a shout outside the window. Looking out, she recognised the walls of Trinity; they were circling College Green. She pressed her nose against her sleeve, blocking out the smell of the streets. Sewage, smoke, urine. The sour fug of hops from the city's breweries. They'd arrived.

The passengers' groans were tinged with relief. It was now twilight; they had been on the road since seven that morning. Through the murky air she saw Dr Stokes waiting amidst the crowd. He raised his cane in greeting. He wore a handsome grey coat, but she noticed immediately that he'd aged: stouter, white threading his whiskers.

'Dear Miss Hutchins, is it you? Pearse, Miss Hutchins's trunk.' A young man, still in the teens, shouldered his way into the crowd.

'You see we have a new Plunkett,' Dr Stokes said.

'I won't know anyone,' she said.

'That will soon be remedied. Re-acquaintance is a quick business, where there was once intimacy. Let's get away from this circus.' He ushered her towards his standing carriage. 'We're still at Harcourt Street, fortunately, and haven't far to go.'

With the carriage doors closed, Ellen hid a sudden shyness by straightening her skirts and pulling at her gloves. Dr Stokes, for whom awkward social moments didn't exist, gazed at her unabashedly. Was he observing the trace of the years across her face, as she did on his? 'The journey is ruinous,' he said, as if confirming her thoughts. 'I have not attempted it since I was a young man.'

'I haven't slept properly since Ballylickey,' she said. 'But then, I sleep poorly at the best of times.'

'You shall rest as long as you want in your old room,' he said.

On Grafton Street, more of the buildings had been given over to shops. From the carriage she saw flashes of every conceivable luxury: silk ribbons and bonnets, cured meats, silverware, jewellery, leather crops, boots, gloves, sweets, cakes in rainbow colours. Merchants waited in the doorways and hawked for business; many were fat, all well-dressed. A seagull swooped from a roof and snatched a pigeon wing from the cobbles; a young woman in crimson silk lifted her skirts and stepped around a mound of horse dung.

'Mr Agardh will call on you tomorrow,' Dr Stokes said.

'You've already met him?'

'He came to dinner. He has little English, so it's difficult to say what manner of man he is. I showed him my herbarium; we managed a dialogue in Latin. My impression was of an intense, serious personage.'

'My Latin will hardly suffice,' Ellen said. 'I know only a few botanical terms, none of which are much use for general conversation.'

He waved his hand. 'Don't concern yourself. I imagine he simply wants to see for himself what a lady botanist looks like. They must not have many in Sweden. Apparently he also intends calling on Mrs Griffiths, in Torquay.'

'I should like to meet Mrs Griffiths myself,' she said. 'Though I can't imagine travelling from Sweden to do so.'

Dr Stokes shrugged. 'No doubt he's an eccentric. But he has an excellent reputation ... ah, we're home.'

Here were the granite steps and the rising wall of brick, butter-yellow through to soft brown. The arch over the front door, the white columns either side. The door, then painted black, was now ivy green. She had walked through it as a frightened girl, with Dr Stokes leading the way, just as he did now. She watched the scene, as if from a remove. Did time snag on such key moments, revived later by memory – moments that linked to form the chain of a life? In that brief flash she was altered, immeasurably. She flexed her hand in its glove. Underneath, the same fingers, muscle, skin, blood.

Her boots were new, made for the journey. She stepped into the hall.

Where had that other girl gone, her innocence, her hopes and particular terrors?

Louise's face also showed the passing of time, her thickened waist the result of the three children she'd borne in the years since. Yet her eyes glittered, unchanged; she examined Ellen as frankly as her husband had. Then she surprised her by stepping forward, placing her hands on Ellen's shoulders, and kissing her on each cheek.

'Isn't that how the French do it?'

Ellen smiled. 'Yes, ma'am.'

'How rosy you are, Miss Hutchins,' she said. 'No doubt you spend a lot of time outdoors.' Her voice, too, the same as it had been: soft and sharp, a needle wrapped in cotton, though Ellen no longer feared its edge.

'I collect plants, as Dr Stokes does,' Ellen said.

'Oh, he has kept me well informed as to your progress. He takes personal pride in your achievements. "Who'd have thought," he often says, "that Miss Hutchins, so unassuming, would become my equal?"'

'I would never presume to think myself Dr Stokes's equal,' Ellen said.

'My dear, of course not.'

'Though in botany, perhaps, I may consider myself at least a colleague of sorts.'

Louise's eyebrows lifted. 'You do? Well, it's true you have surprised me. When I think how timid you were, how you lurked in corners like a wee mouse. Though you were ever watchful, taking everything in, were you not?'

How to answer this? Ellen made a show of looking around the drawing room. 'You've redecorated,' she said.

'Minor changes,' Louise said. 'My hands have been full. Ah! And here's the source of my toil.'

Scuffling at the door: a group of children came in, followed by a young woman Ellen didn't recognise. She wore a nurse-maid's cap, carried a swaddled baby in her arms.

'See if you remember anyone, Miss Hutchins.'

'Can that be Harriet?'

A plump girl with dimpled cheeks. Her hair was neatly dressed, plaited close to her head, in contrast to the other children's free-tumbling locks.

'What age are you now, Harriet?'

'Twelve.'

'It's many years since I saw you. No doubt you've forgotten me.'

The girl opened her mouth:

À Rouen, à Rouen, sur un petit cheval blanc
À Verdun, à Verdun, sur un petit cheval brun
À Cambrai, à Cambrai, sur un petit cheval bai
Revenons au manoir sur un petit cheval noir
Au pas! Au pas!
Au trot! Au trot!
Au galop! Au galop!

Another girl, fair, slender, took Harriet's hand. 'She often sings that rhyme, Miss Hutchins.'

'I remember you would not learn it, Minnie.'

'Harriet was always more clever than I.'

Louise laid her hand on the shoulder of a tall boy, fondly brushing aside his curling red-gold hair. 'And here is Charles. He could only toddle about when you were with us.'

'Charles. You were a delightful child, ever smiling and agreeable.' Charles, robust, very like his mother, tried not to look pleased, though a dimple appeared by the side of his mouth.

'What of me? What was I like, then?'

This from a small boy: large, sparkling eyes, a thin, rigid little form, as though compressed with energy, on tiptoe to assert himself amongst his siblings.

'I confess I didn't know you,' Ellen said. 'You were too young to take much notice of – just a babe in arms when I lived here.'

'Yes,' Louise said. 'This is William, the true genius of the family.'

'Though I am just six,' he said.

Ellen laughed. 'Almost a gentleman.'

He regarded her, his little face solemn. 'You're the lady from Bantry that sends Papa plant specimens.'

'William is also a botanist,' Louise said.

He squirmed away from his mother's hand and cast a fierce look at his giggling siblings. 'I help my Papa,' he said.

'He has much to teach you,' Ellen said.

'We often go to the countryside, to our house there. I prefer it than in town.'

'It's delightful in July,' Louise said. 'Not so much in January.'

'Country life is not easy once winter descends,' Ellen said.

'The cold doesn't bother me,' William said. 'It's Mama and the girls who complain. And the servants. They're a feeble lot.'

'Enough,' said Louise. 'Off to bed with you, now, all of you.'

William hung back. 'Miss Hutchins,' he said quickly. 'May I show you my plant collection tomorrow?'

'Don't plague Miss Hutchins, Willy.'

'I should be glad to see it,' Ellen said.

'You see, Mother?'

'Yes, well. To bed, now.' William closed the door behind him. Louise sighed. 'You see I now have two in the family. Truly, there is no escaping botany in this house.' She tilted her head, regarded Ellen. 'Speaking of botanists – you could have called on Mr and Mrs Taylor while you are here but I believe they are away in England.'

A knot unclenched itself from under Ellen's breastbone. 'A pity,' she said.

'Does he still visit Inchilough?'

'Not as often, I hear.'

'Ah.'

In the hanging pause, Ellen resolved to change the subject. 'Is Dunne no longer with you?'

Louise wrinkled her forehead. 'Dunne?'

'The nursemaid who was here at the time I was.'

'Ah. Yes, Dunne. As it happens, I ended her employment not long after you left. She turned out to be unsuitable.'

'Oh?' Ellen had a flash of memory: poison spun through a little girl's mouth. The crimson glow of a glass of wine held against the sun.

Louise's mouth curled. 'She had a loose tongue, and was prone to tittle-tattle. Untrustworthy, in the nursery and anywhere else.'

'I see.' She waited, but Louise said no more.

◈

Mr Agardh arrived early; Ellen had to rush from the breakfast table.

She had supposed he would be tall and fair. Instead a short, dark man stood by the window, dressed in black, with a white collar like a clergyman. Dr Stokes urged her forward, made the introductions. Mr Agardh clasped a ledger tightly under his arm. His hair was short, black and wiry, like badger fur. His nostrils quivered.

'Miss Hutchins,' he said, adding extra sibilance to the words. He held out his hand, shook hers vigorously. 'An honour,' he said. 'I come from far – ' he waved his other hand towards the window, presumably to indicate the sea, ' – to meet you.' Dropping her hand, he opened the ledger and carefully flipped through the pages. Straight to business, she thought. He smiled, showing narrow, sharp teeth, and pointed. She bent to look.

'*Leiocolea bantriensis*,' she read aloud. She smiled. 'Yes, Mr Agardh. Bantry Notchwort.'

He stood on his toes, danced a little back and forth, nodded. 'Ah!' He flipped the pages forward and stopped again.

'*Thelotrema isidiodes*,' he said. 'Yes?'

'This too,' she said. She felt a strong desire to laugh. Instead she kept smiling, trying to seem natural. Her cheeks began to ache.

He spoke in his own language, a mixture of guttural and whistling sounds. She held out her hands. 'I don't understand.'

The Swede blew air out of his cheeks, shook his head. He stepped back and held himself straight, placed his hand on his breastbone. 'Carl Agardh,' he declared. He held up both hands, then a single hand. Emphatically, he laid the hand again on his breastbone.

'I believe he's telling us his age,' Dr Stokes said. 'Twenty-five.'

He looked at Dr Stokes, indicated Ellen.

'Ah,' Dr Stokes said. 'He asks what age you are, Miss Hutchins.' Using her fingers, she showed him – two hands, twenty. One hand, five. 'Also, twenty-five.' She would have found it difficult to guess his age. He had the gravity of an older man, the innocent eyes of one much younger.

'Mr Agardh seems astonished at all you've accomplished, and why would he not be?' Dr Stokes said.

The door opened. Servants, with trays: tea.

Despite Dr Stokes gently pushing him forwards, Mr Agardh wouldn't move. 'Come, sir. Tea for a damp Irish morning.' Ellen stepped towards the sofa, thinking he might follow.

A large globe rested on a mahogany stand, set between two chairs. Ellen had no memory of it: a varnished sphere,

delicately coloured. A new purchase, part of the improve-
ments. Mr Agardh appeared to notice it for the first time. He
walked to it and laid his hands on the smooth circumference,
adjusting the position until he found what he was looking for.
Turning back to Ellen he asked, 'You?' She walked back to
join him. His finger prodded.

'Yes, Bantry,' she said. 'And Lund? Where is Lund?'

'Ach, Lund.' He spun the globe to the left, located Scandinavia.

'The sea?' she asked. 'Close to the sea?' He pinched his
thumb and forefinger together: very close. His impish face
serious now, hoping she understood.

He wanted to reach her, to know her. He had travelled
across cold seas and on dangerous roads, put himself at risk.
She was glad she'd come. He offered her his arm; they walked
together to the table where tea had been set out. She loomed
over him, she could carry him away in her pocket.

If she had language to do so, she would say: *Dear Mr
Agardh – with one hand I hold a knife, a paintbrush, I adjust the
microscope lens. With the other I wipe my brother's face, help my
mother to the chamber pot, darn moth holes and feel for the pulse
of a newly laid egg.*

For the next fortnight she rejoined the family in their daily
routines: Dr Stokes's comings and goings, the children's les-
sons and play, Louise's social rounds, mealtimes. William
showed her his herbarium. She studied the pages; nodding at
his chatter, keen to show him proper attention. He spoke flu-
idly, with confidence. Already he displayed an understanding
of the subject she recognised as exceptional.

'Where do you go to school, William?'

'I don't, I have a tutor. Otherwise I accompany Papa on
his ramblings, or help in the laboratory.' Dr Stokes appeared
to favour him with unusual time and energy. The boy clearly
benefited from it, as a young plant dug into fertilised soil.

'How fortunate you are, to have such a father.'

'I know it.' The boy shrugged: he appreciated it, but truly, was it not his due? 'Miss Hutchins, shall I recite something for you?' He scarcely waited for her nod before launching into verse. His eyes glazed over, as though focused on some internal vision.

It fell about the Lammas tide,
When husbandmen do win their hay,
Earl Douglas is to the English woods
And a' with him to fetch a prey.
He has chosen the Lindsays light,
With them the gallant Gordons gay;
The Earl of Fife, withouten strife,
And Sir Hugh the Montgomery, upon a gray.
They have harried Northumberland,
And sae have they Bambro'shire;
The Otterdale they have burned it haill,
And set it a' in a blaze of fire.

He waved and thrusted his arms; his pitch rose to a squeaking excitement. At the final verse he slowed and assumed a tone of great sorrow, comical in its gravity. Four-year-old Gabriel had crept into the drawing room and lay at his brother's feet watching the performance, mouth open, eyes wide.

… This deed was done at Otterbourne,
About the breaking of the day;
Earl Douglas was buried at the braken bush,
And the Percy led captive away.

He bowed neatly.

'Well done,' Ellen said, clapping. 'Most impressive.'

'Shall I recite another? I know all of Scott's ballads by heart.'

'I don't doubt it. Tomorrow, perhaps. Pleasure is greatly increased by anticipation.'

She ventured into Archer's booksellers on Dame Street and bought Walter Scott's newly published *The Lady of the Lake* (Dawson Turner had recommended it). Stepping back out onto the dusty street, she recognised a familiar figure, distinct amongst the jostling crowd of the pavement. Tom Taylor. He must have returned from England. Instinctively, she raised her hand, before noticing that a woman walked close alongside him, clutching his arm.

Ellen shrank back into the shade of Archer's shop front. As Sam had said, Mrs Taylor was small and slight; Ellen caught a glimpse of black hair, dark eyes, a pretty, peevish face. She turned away and stared into the display of books in the window until the couple's blurred reflections melted into the throng.

Once back in Harcourt Street, she stowed *The Lady of the Lake* carefully in her trunk. It would fill some of the long hours once back home.

She woke in the early hours of the next morning with a dull pain in her stomach. Her bleed was so erratic she could never predict it; it disappeared altogether for months at a time. Now it appeared it was on its way. Thankfully she'd thought to pack clean rags. She lay awake until dawn, fighting off the twisted thoughts that always intensified at this time, scolding herself to calm. Despite the softness of Louise's linen, she longed for her own bed at Ballylickey.

After questioning the reason for her pale face, Louise offered to send out for camomile. Ellen obediently drank several cups of the weak yellow tea (though it smelled faintly of cat urine); the only effect was to send her repeatedly to her room to use the chamber pot. Still no sign of the telltale spotting. That night she asked for a hot brick, and huddled it to her stomach until she fell into a fitful sleep.

'Have the cramps eased?' Louise asked, eyeing her at breakfast.

'They are not so troublesome as they were,' Ellen said.

'Should I ask Dr Stokes to examine you?'

'No,' Ellen said, quickly. 'They will pass, I'm sure.'

Louise laid her hand on Ellen's arm. 'We women are cursed. By that, and a hundred other torments. I am plagued constantly, myself. Of course, once you have children, you are never free of pain, for ever after.'

'I've had a feeling for days now that something is amiss.'

'Amiss? In what way?'

'I can't say.'

'I think you're homesick, my dear. It's natural, so far from your mother.'

As always, the table was placed too close to the fire. Ellen had become used to the chilly rooms at Ballylickey. 'May I raise the sash for a moment?' she asked.

'Are you faint? I can have vinegar brought from the kitchen.' Louise called the manservant. 'Pearse, open the window and let in some air.'

Ellen stood in the draught. Her pulse slowed.

A woman clad in rags shuffled along the path opposite, shoving her hand at passing citizens.

According to Louise, the society ladies she took tea with complained incessantly about the crippled and the maimed, the destitute beggars, desperate mothers and god-forsaken children, how they spoiled Dublin by cluttering the thoroughfares with their wretchedness.

'Their lips quiver in outrage,' Louise said. 'They would condemn the entire poor to the workhouse.'

'Idle, foolish women,' Dr Stokes said. 'They have no pity, or shame.'

'What is the answer, though?' Louise asked.

'I have none, save education, decent housing, access to better-quality food. But your ladies would prefer to sweep the problem from the streets, hide it from view behind high walls.'

The woman had collapsed to her knees. Impossible to say what age she might be, her face masked by dirt, over-hung with matted strands of hair. A gentleman in a blue coat swerved from her grasping fingers without breaking stride, his face twisted in disgust.

'Miss Hutchins!'

She turned from the window. Louise stared at her. 'I said, there's a letter come for you.'

She mumbled an apology, took the letter. Arthur's hand-writing. She tore open the seal and shook out the half sheet.

Ellen,

Mother has taken a turn and is fast declining. She was found in her bed yesterday morning, unable to speak or move. You are needed urgently. Arrange to travel as soon as you possibly can.

Arthur

Dated two days before.

Louise came across the room, took her hand. 'You look stricken, Miss Hutchins.'

'My Mother has taken ill,' she said. 'I will leave today, if I can.'

'How unfortunate.'

'If a servant could help me pack my trunk ... and if Dr Stokes has not yet left I would ask him to take me to the coach stop.'

Louise shook her head. 'The bond you must have with your poor mother.'

At the steps of the coach, Dr Stokes took her hand. 'Take care of your health,' he said. 'The glitter in your eyes shows over-excitement. Remember to take enough nourishment.'

'I try, sir,' she said. 'As you remember, I never had much appetite.'

'Dear child. You have such will, so much quiet energy, I doubt the whole of Mrs Stokes's pantry could fuel it.' He increased the pressure on her hand for a moment, released it. 'You didn't get to see Mackay, or his new botanic garden.'

'I hoped I could stay longer.'

'We would keep you with us always.' He sighed. 'In times such as these, parting with a friend brings inevitable fears for when one might see them again.' Looking into her face, he added hastily, 'I predict Mrs Hutchins will have rallied at the prospect of your return. It sounds to me like a mild case of apoplexy. Though you do right to go to her side.'

The women whispered to one another, their words clearly audible in the confined space.

'She has neither style nor good taste.'

'If nature has not provided them, money cannot acquire either.' By the time the coach reached the plains of Kildare, they had thoroughly dissected their subject; the unknown woman had shabby clothes, an uncultured husband, ugly children, lazy servants, an ill-behaved dog. At last, unable to stop herself, Ellen leaned forward.

'Have you perhaps another topic of conversation?' They stared at her, open-mouthed. 'I already know more about your friend – if that is indeed what she is – than I like to. It's a long journey, could we not all rest for a time?'

'How unspeakably rude!'

She shut her eyes, not caring to see their outraged faces, and pictured the waterfall, imagining the moment she would feel the mist on her face, the utterly clean, pure breath as water hit rock.

Dear child. She had wanted, longed, to be Dr Stokes's child, envied his children when he'd kissed them or called them pet names. Yet he had never called her 'dear child' then, not once. Nor, she reminded herself, had she been one. Eighteen, a full-grown woman. She blushed when she thought of her secret thumb sucking. Longing to lay her head on Louise's breast for petting, as Harriet did.

Had she not also envied Louise? Secure as a nut in the shell, queen of her household and treated as such by her husband. The glimpses of intimacy between them had seared into her mind, resurrected many times in the dark.

Dr Stokes had shifted ground in her opinion. She had matured while he had shrunk. Her awe had lessened, though her gratitude remained unaltered.

Although he had saved her life, the road he'd set her on had led to yearnings that couldn't be satisfied, success with limits. She couldn't blame him for that. He had not made her his child; he had never assumed her anything but capable, setting the tone for Mr Mackay to follow. And from Mr Mackay to Mr Turner.

TWENTY-THREE

Saliva drooled from Leonora's mouth. Her eyelids flickered, her lips smacked like a baby's. Ellen gently wiped her chin.

'She seems more alert,' Jack said. 'Her colour is better.'

Ellen took Leonora's immobile fingers in her own and stroked the back of her hand until her eyes closed. 'She managed a sup of broth.' The spill-stained sheet would have to be changed later. Kate had done her best, but even so the tang of faeces and urine lingered, like a foul taste in the mouth.

'Will she talk again?' Jack said.

'God knows.' A sudden thought: 'Should we ask Reverend Smith to come?'

'That fool?' he snorted. 'To mumble prayers over her head, like the Catholics do – "last rites", as they call it? We've no need of that.' In Jack's mind, Leonora, fragile as glass though she might be, had been there as long as he could remember, would live forever.

'Mother might be comforted by seeing him.'

'Aye, or be terrified, to wake and see that beaky humbug by her bedside.'

'She likes him. Or at least, she respects him.'

'I couldn't listen to his sycophantic wheedling. Not now.'

Leonora hiccupped, her face slack. 'Perhaps you're right,' Ellen said. 'We should pray ourselves.'

'You do it,' he said. 'God does not listen to me.' Still, as she prayed she saw his lips tremble and move, following her own whispered entreaties.

❧

She slit along a new page of *The Lady of the Lake*. No stir from the figure in the bed. The logs hissed and cracked in the grate; a stray twig flared and burned in a yellow glare. Her deformed shadow reared against the wall. Four o'clock in the morning: her eyes burned.

A grunt from the bed. The hand twitched on the counterpane. Ellen shook the rug off her knees and went to the bedside.

'Are you thirsty, little mother?' Squeaking, like that of an injured rabbit. She tilted Leonora's chin upwards and spooned water into her mouth. Leonora's pale lashes flickered alarm.

'There. That's better, isn't it?' She tried to keep the tone of a mother to child from her voice, failed. In truth, their roles had reversed long since. She pulled at the blankets, tucked them around Leonora's shoulders. 'I'll read to you for a time, it might soothe you.' Before she sat again she poked at the fire, sending a storm of sparks dancing into the air. She dragged her chair closer to the bed, cleared her throat.

As died the sounds upon the tide,
The shallop reached the main-land side,
And ere his onward way he took,
The Stranger cast a lingering look,
Where easily his eye might reach
The Harper on the islet beach,
Reclined against a blighted tree,
As wasted, gray, and worn as he ...

The fire shrank, the candle spluttered downward.

To minstrel meditation given,
His reverend brow was raised to heaven,
As from the rising sun to claim
A sparkle of inspiring flame.
His hand, reclined upon the wire,

Seemed watching the awakening fire;
So still he sate, as those who wait
Till judgement speak the doom of fate;
So still, as life itself were fled,
In the last sound his harp has sped.

She closed the book. Her hairpins tugged against her scalp. Oh, to take her hair down, relieve the pressure on her skull! Hairpins and stays: all that held her together. She cursed them both. Leonora's head slumped to the side, but her chest continued to rise and fall. Ellen allowed her own eyes to close for a moment, her body to sink into the chair. When she next woke, dawn light had slid across the floorboards. Deep silence: not even the early screams of birds.

Something stood by her mother's bed.

Its hair and skin were grey, as though sifted with ash. This is a dream, Ellen thought, and it cannot hurt me.

'Katherine?'

The hands fluttered, broke apart like charred paper. Movement, closer to the bed. Ellen sat up.

'Leave her!'

A shimmering, out of focus, as if viewed through an ill-adjusted microscope, then it vanished. Ellen had frozen in the chair, her heart flailing. After a minute her pulse slowed. Her eyelids grew heavy; her thoughts darted away beyond her grasp, like minnows in a rock pool.

She slept until woken by a clatter: crockery, rattling on a tray. Kate's head came around the door. 'There you are,' she said in a low voice. 'When I saw your bed empty, I knew you'd be in here.'

Ellen shivered, stretched, yawned. Kate placed the tray on a side table, poured tea. Ellen stared at the steaming amber liquid.

'Have you sat there all night? And the fire gone out.' Kate set the pot down and went to the fire. She poked at the log, now a blackened carcass.

'I read to Mother for a while.' Ellen took the cup in her hands and blew on the surface of the tea.

'The sheets want changing. We'll have to move her to the chair. Will I bring up the stirabout?'

Leonora's mouth jerked.

'Let's sit her up,' Kate said.

'How is she, do you think?' Ellen asked.

'Ah, she'll be about Bantry again one of these days.' Kate's furtive eyes told the truth: Leonora might never make it downstairs, go to Bantry or anywhere else.

In which case Ellen would be nursing her for the rest of her life.

'Are you alright, Miss?' Kate said. 'You look peaky.'

'I'm fine.'

'The last thing we need is for you to collapse.'

'There's no question of that,' Ellen said, standing up. 'A dream I had last night, that's all.'

'What dream?' Kate asked, then immediately holding out her hands as if to ward off an attacker: 'No, don't speak of it. It'll only bring bad luck.'

Ellen lifted *The Lady of the Lake* from the table. 'Too much Walter Scott, I fear.'

'No wonder you can't sleep in your bed, like a normal person. I told you reading was bad for you.' Kate flapped at the book. 'Haven't you enough to worry about without scaring yourself with stories?'

'You like a tale yourself, Kate, around the fire.'

'I'm not high strung, like some.'

'High strung. Is that what I am?' A fine thread, about to snap.

❧

She came in warm from her walk; only her hands were cold. Her fingers itched as she held them over the heat.

'Master Arthur is here, Miss,' Kate said. 'He's waiting in the parlour.'

She turned from the fire. 'We weren't expecting him.'

'No, Miss. He rode over early. Will he take breakfast with you?'

'I don't know,' she said, hurriedly unwrapping her shawl and smoothing her hair. 'I'll ask him.'

In the parlour, he stood facing the fire, one boot on the edge of an andiron. He looked better these days, stronger. He had put on weight, it suited him.

'Have you come to see Mother?' she asked. 'She's much the same as last week.'

'Shut the door, Ellen. Have you heard from Emanuel?'

'Not for some time.'

'I have had a letter.' He took it from his inside coat pocket.

'I see.'

He indicated a chair. 'Let's sit.'

'Will you want breakfast?'

'Later,' he said. 'Sit, Ellen.' She took the chair opposite him. He opened his mouth, closed it again, frowned, as though considering his words. 'It might come to nothing, but I wanted to make you aware of Emanuel's latest scheme.'

'Yes?'

'He suggests coming home. To live at Ballylickey.'

She blinked at him, incredulous. 'He has hardly set foot here in years.'

'Nevertheless, he says he is giving serious consideration to moving back.' He threw out his hand, as though casting something away. 'It puts me in an exceedingly awkward position with the lease. At best, the title is unclear. If he has plans for Ballylickey, makes a move, I may have to challenge him.'

'What sort of plans?'

'Who knows? He possibly doesn't know himself, exactly. Which leads me to suspect the worst.'

'Of course, Mother would love to have him home again.'

Arthur frowned. 'She may come to regret it.'

'What can you mean?'

He didn't reply, stared moodily into the smoking fire. The logs were damp. He sniffed, took out his handkerchief, used it to blow his nose, tucked it away. 'He's suggested it before, over the years. Hints, ramblings. I thought I had successfully dampened them down. And Samuel always managed to divert him. But I now fear his circumstances are such that Ballylickey presents a solution to many of his problems.'

'What problems?'

'Money, presumably. And he claims an attachment to the place, which may well be true, though as you say, he has not demonstrated such in recent years.'

His words solidified, took on weight. 'When do you think this might happen? Next month? Next year?'

'He doesn't go so far as that. At present it's all speculation.'

Silence. The clock on the mantel ticked steadily.

'Will you write back to him?' she asked. 'Endeavour to discover his motives?'

'Certainly.'

'Should I mention it when next I write?'

'No. There's little point. Stay out of it.' He paused. 'I doubt I can influence him by any method other than threat.'

'Oh, Arthur, pray do not start a fight with Emanuel!'

'It isn't a question of "fighting", Ellen. We're not in the schoolyard. Brothers should not quarrel, I agree. It leads to fragmenting of the family. But I must quash any schemes he has in this direction, once and for all.'

'What of Mother? Will you tell her?'

'Would she absorb it?'

'She knows all that goes on,' Ellen said. 'I'm sure of it.'

'Well, with luck, I can resolve the issue – by whatever means necessary. She need never hear of the matter.'

'Jack has heard hints of a quarrel before, the same as I have.'

He thought for a moment. 'I will find the right time to talk to Jack.'

A knock on the door and Kate came in, head down, eyes lowered. What had she heard? Ellen didn't begrudge her listening at doors. Anything affecting the family affected her and the other servants. Fruitless to try and stop her, at any rate.

'Master Jack is up, sir, and asks that you join him in the dining room. He's most anxious to hear the reason for your visit.'

'Just passing by, Kate, on my way to Bantry from Ardnagashel, and had a craving for one of your excellent breakfasts.' Arthur stood. 'We will speak later, Ellen. Have faith in my judgement.'

'Of course, Arthur,' she said.

She caught a sharp glance from Kate before they went out the door, but nodded back as though all were well. An ordinary morning, at the end of the world.

Prosperous, Co. Kildare
February 1811

Dearest Ellen,

I hope your mother has recovered. I have prayed for her, and you.

Once again I am running to my chamber pot in the mornings; the physician confirmed today what I already knew, that I am once again with child. Mr Aylmer would no doubt like a boy though I would prefer another daughter, girls are such a comfort. Boys leave you, to find wives of their own and develop new loyalties. A girl remains close even if she marries, and of course, she may not marry at all but stay with her mother, as you have done.

You write so seldom now that I take it you are happy and busy. Though our lives are very different, I believe something remains in each of us of the young girls we were. I still occasionally play with my old fan and preen into the mirror, when no one is looking, though I am old and married, and expected to be dull. Who does it harm?

Since my father's death I feel curiously light, though it's true I cried fit to break for six months after. I woke one morning and realised that the darkness I feared would surround me forever had fled. It was never my mother that haunted me, she was only ever a blessed presence.

I say this to bring you comfort, dear Ellen, though I scarcely know what I mean. We can be truly free, in our minds and souls. The old passes away, we make ourselves anew.

Your little watercolour sits in front of me as I write these lines.

Yours ever affectionately,
Caroline Aylmer

YARMOUTH, ENGLAND
JULY 1811

TWENTY-FOUR

'Have you fixed on a name, sir?' Mr Noakes, the accoucheur, lifted his hands from the basin of pink tinted water and grabbed the clean cloth the nurse held out.

Mary lay against the pillow dressed only in her shift, eyes half closed. Strands of hair sweated to her forehead and temples. Dawson bent over the basket. His new daughter stared somewhere beyond him, grunting softly. Her fingers, damp and biting, coiled around his.

'She is to be called Ellen,' Dawson said, 'After a family friend.' Mary nodded and strained at a smile.

'Miss Ellen Turner,' Noakes said. 'As sturdy a child as I've delivered. She should thrive.' The bedsprings creaked as Mary shifted position. Noakes extended his arm towards the door. 'Come, sir. Let us leave the nurse to tend to Mrs Turner.'

Before he left, Dawson kissed Mary's hot forehead. 'Well done, dearest,' he said.

She clutched his hand, as the baby had. 'You're disappointed,' she said.

He swallowed. 'How could I be?' She rested her head against his chest, breathing heavily. 'Rest now. You're spent.' Obediently, she let go of his hand and leaned back, already drifting into sleep. The nurse hovered by the bed, impatient to draw across a veil of feminine secrecy between he and his wife. There would be mysterious interventions under bedclothes, whispered consultations, a language of nods and half-words designed to be indecipherable to her husband of fifteen years, with whom she had shared moments of the

most extreme intimacy ... He quickly followed the accoucheur out of the room.

After seeing Noakes out – pressing a coin awkwardly into
his hand – he climbed the stairs again, went into the drawing
room. The younger children played quietly on the rug. The
older three sat on the sofa, holding hands. They stood as he
came into the room.

'You have a new sister,' he announced. They released their
hands, and began jabbering at once.

'Does this mean we can go out now?'

'Can we see her?

'Will Mama come down?'

'Mama is tired and must sleep,' he said. Maria came to
stand beside him, worked her hand into his.

'Are you pleased, Papa?'

'Very.'

'Truly?'

He smiled down at her solemn face. 'Truly. She is a flower,
more than I deserve. A blessing, as each of you is a blessing.'

The corners of her eyes crinkled. 'Then we are pleased
also. Even if she is not a boy.'

BALLYLICKEY, CORK
AUGUST 1811

TWENTY-FIVE

She would think of herself from here on as her little name-sake's particular guardian. In spirit, at least.

Did she resemble her mother, or father? It was generally preferable for girls to resemble their mothers, though she knew of some who shared their father's features to no disadvantage. It was also desirable for boys to take after their fathers, though for different reasons. Each of her brothers was a Hutchins, through and through: tall, light-haired, sharp-featured. This made their enmity the more distressing, as though they turned on themselves. Impossible to strike at those closest to us without feeling the force of the blow.

She wished the infant an easy happiness, one she had found elusive. Better if baby Ellen grew to have lightness of spirit, so attractive in a personality, young or old. Her own character had been formed by the death of her father, the years of isolation at school. Though she sometimes thought of this person, shivering and weary, as naught but a skin, that she could cast off and let fly in the wind, like a length of dried seaweed whipped along the shore.

Mr Turner recognised her, this happier woman within, encouraged her forward. Was this not friendship, the recognition of one's finer self in another's eyes?

He wished she could be godmother to his child. This brought her joy, but also a spike of sorrow. Perverse: her every emotion was wrapped in another, as the layers of an onion.

Only when writing to Mr Turner, did she now approach the core.

TWENTY-SIX

The spring, after having continued for an hour and half spouting its waters in so lofty a column, and with such amazing force, experienced an evident diminution in the geyser's strength; and, during the space of the succeeding half hour, the height of the spout varied, as we supposed, from twenty to fifty feet; the fountain gradually becoming more and more exhausted, and sometimes remaining still for a few minutes, after which it again feebly raised its waters to the height of not more than from two to ten feet, till, at the expiration of two hours and a half from the commencement of the eruption, it ceased to play, and the water sunk into the pipe to the depth of about twenty feet, and there continued to boil for some time.

A gift, courtesy of Captain Mangin, carried across the Irish Sea: Mr Hooker's newly published *Journal of a Tour of Iceland*. Before she cut the string of the parcel, she sniffed at the wrapping, fancying a whiff of brine and tar. The open sea: whales and other monsters, twisting in the dark, the oncoming loom of unknown coastlines.

She had been wrong in her imaginings about Iceland. According to Mr Hooker, rather than being dormant, encased in ice, it pulsed and boiled. The earth there cracked open in a state of constant creation.

During Mr Hooker's voyage home a fire onboard had destroyed most of his notes, drawings and journals, so that much of his account had been written from memory alone. How observant he was, how thorough his recall! Plants, reported in exact detail – *Andromeda hypnoides, Rhodiola rosea,*

Lycopodium annotinum, Orchis hyperborea. Descriptions of food, such as a feast of many courses in a wealthy Icelander's house: terns' eggs, salmon, soup made of sago, claret and raisins, mutton roasted with sorrel. His many adventures – swimming in springs heated by the earth's core (his poodle, brought from England, ran innocently into a scalding pool and had to be carried, like a child, for many days after to save its blistered paws); watching a pale yellow *Aurora Borealis* pulsate in the southern, eastern and western skies.

Enough wonders to amuse for a month. Yet she kept turning back to the frontispiece, the illustration of an Icelandic lady in her wedding dress. Hooker had actually managed to bring the dress back to England, rescued from the fire by the ship's captain. What a costume to wear to your wedding, as though dressed for going into battle! An ornate, high-necked black jacket, moulded to the waist and finished in gold braid, resembling something an officer might wear on parade. The white headdress, embroidered in gold, plumed two feet upwards from the lady's head, like the crest of an exotic bird. A crimson ornament hung around her neck. She looked gallant as a ship and fierce, neither demure nor modest as a bride was supposed to look.

Did such clothing inspire courage, or allow you ape a courage you didn't feel?

She grasped a handful of material from her skirt, and inspected it: food stains, dirt so ingrained it could never be fully washed out. Her skin itched; she felt a restriction under her arms, around her breasts, down her sides. She pulled at the neckline, trying to loosen it from her body.

Soon she'd dry out completely, as fusty as her plant samples. No need for battle dress, or even plain white muslin.

The crooked walls of the old house loomed inward, its corridors choked with shadows. Dust lay on every surface, in her mouth. She needed air. Jack's voice called from his room, 'Ellen, is that you?' A pile of stinking bed linen lay at the foot of the stairs; she'd earlier helped haul it down. Passing, she kicked it out of the way.

She should go to Jack, see what he needed. Instead she went through the kitchen and snatched her coat, bonnet and gloves at the back door.

Outside she shouted, 'Dan!' After a minute he came, wiping his hands on his trousers.

'Miss?'

'Bring the gig around.'

'I can do an errand for you once I've finished the whitewashing, Miss.'

'I want to go myself.'

'What is it you're needing?'

'I don't know, but I'll think of something.' He looked towards the house, as if hoping Kate might appear. She persisted. 'I only have a little time before my mother wakes up.' Still he dawdled. 'Dan, do as I ask!'

She waited for him to hold the horse steady, then climbed on, clicked the animal forward. 'Get up!'

Scent of honeysuckle and wild rose in the hedges, air streaming past her face. Small stones spun away from under the wheels as she gave the horse his head. After a few minutes she pulled on the reins, restraining him, and let her breathing subside and match the slowing clip of hooves.

What had got into her? A surging in her blood that she couldn't control. It happened more and more lately.

In Bantry she left the gig by the deserted quayside. Most of the boats were still at sea; the few that remained listed with the ebb of the water. Yelping gulls, the dull slap of water against wood. Two women in bare feet, their skirts hitched up, went past carrying baskets of sand on their backs. On the quayside, a seagull stabbed at the shell of a crab.

The town was livelier, with people milling about the streets. She murmured hello to those she knew. Some stopped, grasping her arm and enquiring after Leonora. She told each of them, briefly: my mother is confined to the house for the immediate future. Their clucks of sympathy followed her as she hurried away.

In the churchyard an old man cut at the grass with a scythe. He lifted his hat and nodded at her civilly. She bade him good day and walked past. He bent again, wide-legged, to his work, mowing swathes left and right. Poppies, daises and cornflowers fell with the grass: a pity.

She walked through the rows of Trenwiths, Lavers, Swantons and Warners. Names softening with time and rain. How many secrets had been spoken against the rough stone, to these old companions – they issued no judgements or recriminations. Only silent comfort, constancy. At the end of the row, she laid her hand on her father's plain granite headstone. Modest, for a man of his standing, just his name, date of birth and death. No mention of his career as magistrate, importer, landlord. Below that: his daughter Katherine, 1765–1789. The other children and infants appeared to have been buried elsewhere, or hadn't been added to the stone. Space had been left for Leonora's name, and the names of those who would follow her into the ground.

Ellen kneeled on the soft earth. The churchyard promised a kind repose. Though she would choose a different resting place.

The woods. Lie down, return to the dank soil with the half-light filtering through the trees and the sweet, heedless chatter of the birds.

Or the dark, deep sea. Carried out on a quiet tide.

The quickest way was off the cliff. Men carrying torches would search the headlands, shouting her name. Yet she wouldn't want her body splayed on the rocks for the gulls and crabs. Found with her skirts blooming around a tangle of limbs.

Foolish fantasies. Indulgent, sinful. You fell into them all the same, like a bed of feathers.

The swishing had stopped. The mower leaned back his head, tipping beer from a golden bottle into his mouth. She licked her lips and swallowed, suddenly thirsty. The sun rose higher, whiter. Thick fumes from the meadow hay cloyed in her nostrils and throat.

Had you lived, Father, what then? Had you not left us, rudderless, at the mercy of every wind and swell?

Someone else had come into the graveyard, a woman dressed in black, dragging a child by the hand. Mrs Dickson. Ellen stood up, brushing clay from her skirts. Passing, she bent her head and avoided the widow's stricken eyes. The boy escaped his mother's grip and dashed away after butterflies. 'Watch the blade, young sir,' the old man cautioned.

Edward Dickson lay in his grave, a month dead, from poisoning of the blood. They said his head had swelled twice the size, and turned black.

From a distance she saw what she thought was a pile of discarded rubbish – driftwood, branches, discarded planks, propped together and crudely patched with rags and peat. As she came closer, she discerned a creature amidst the debris, muffled in a cloak. Its skinny arms protruded, winding wool onto a spindle. The creature's back distended into a solid hump, so large it seemed impossible that it could sit upright without toppling backwards. Yet it was a young woman's face that peered at Ellen from the cloak's hollow. She held out her hand. Ellen groped about her skirts for a coin, found none.

From within the hump another hand emerged. Claw-like, it scrabbled to free itself. Ellen's heart contracted.

She has someone on her back.

A pair of staring eyes in a brown face, wizened as a January apple. An old woman. Even with her puckered toothless mouth, she had the same features as her daughter.

No one spoke. Then, 'I'm sorry,' Ellen said, backing away, lowering her head.

On the corner a woman stood in a cabin doorway, sweeping dirt on to the road. She might speak English. 'Good day,' Ellen said.

The woman looked up. Behind her, in the gloom, Ellen could see a rush chair, an open fire.

'Good day, Miss.'

'Who are those people?' Ellen asked, pointing.

The woman peered back at the ragged figures. 'McCarthy is their name, Miss,' she said. 'They're from Doonour. They've no men living, God help them. The mother's crippled and can't be left. When the daughter comes to town, she brings her with her. The daughter sells wool, though recently she's taken to begging, poor cratur'.'

'Not from Doonour with her mother on her back? It's ten miles at least ...'

'How else is she to get to town? Not everyone has a horse, or even a donkey. It's walk and break your back, or starve out there in the middle of nowhere.' She jerked her chin, shook her head, beat the broom of twigs against the ground. Dust rose between them. Ellen covered her mouth. 'Have ye not got the likes of them out at Ballylickey, Miss Hutchins?' The woman's tone was familiar, though her eyes had hardened. Had Ellen met her before? No, she thought not, though she herself had obviously been recognised ... 'Some are better served in this life than others,' the woman went on, nodding her head as though at an acknowledged truth. 'That's the way of it.' She resumed her sweeping, stabbing savagely at the cabin's earthen floor.

Ellen pulled on the reins and the horse trotted to a standstill. Kate came outside, shielding her eyes against the evening sun.

'What possessed you to tear off like that, Miss?' she said,

'I needed ink,' Ellen said. 'But I forgot to bring money.' She climbed down.

'Sure they would have let you settle later.' Kate twisted the corner of her apron.

'I didn't think of that.'

'Did you need it in such a hurry?'

'I want to write some letters later.' She pulled off her bonnet, walked past Kate into the house. 'Is everything all right upstairs?'

'I think so.'

Leonora opened her eyes as she came in. 'Where. Were. You?' she gasped.

Ellen put her mouth to her mother's ear. 'Bantry.'

'Ah. Kate. Didn't. Know.'

'I'm sorry.' She brushed crumbs and a stray hair from the sheet, took the water jug and filled Leonora's glass. 'I saw something peculiar there. A beggar woman, and her mother. She walks every day from Doonour with the old woman on her back. I thought it pitiful.' Leonora nodded, smiling. She misheard, no matter how clearly or slowly Ellen spoke. As her speech had come back, her hearing had faded, as though one had replaced the other.

Kate brought in the emptied chamber pot and pushed it under the bed with a clatter. She stood, hands on hips. 'I can't make her understand me at all. And the master has no patience. You're her ears now, Miss. And the only voice she hears in the world.'

With her fingers, Ellen combed wisps of hair back from her mother's temples, and tucked them under her cap. 'The quiet in this house grows,' she said. 'No music, nor laughter, except when Arthur's children are here.' She grasped Leonora's hand. 'We're sinking into silence, one way or another.'

'You're maudlin today, Miss. Have you your woman's pains?'

'No.' They'd finally come, and finished a week ago.

'There's a full moon tonight. Maybe that has you out of sorts.'

'What nonsense you talk, Kate.'

'Aye, and I'll keep talking. As should you. To beat back the silence.'

❦

Christmas had been put away, and the almanac in the kitchen turned over to 1812. A letter arrived for Ellen addressed in an unknown hand.

Prosperous, Co. Kildare
January 1812

Miss Hutchins,

With deep sorrow I write to inform you of the death of my wife, Caroline, a month since.

After a short period during which she suffered a soreness in her limbs and a rash on her skin, she rapidly descended into a fever from which no physician could save her – though I brought in every local man and finally went myself to fetch Dr Lincoln from Dublin.

I confess I might not have thought to inform you, but my wife, fearing the worst and too weak to sit up and write herself, begged me to remember her to you, the greatest friend of her schooldays.

Little Charlotte – Caroline's own image – and the baby, Robert, are some comfort to me.

Your obliging servant,
George Aylmer

Ellen wouldn't have credited him with such kindness. This only added to her shame.

She reread the letter twice, folded it and laid it at the bottom of her correspondence box. She no longer hid the key. Joanna couldn't read; neither Jack nor Leonora were fit to pry. Kate knew all anyway.

She stayed in her nightdress all day, trailing a blanket around her shoulders down the hall to check on Leonora before retreating back to her own shuttered bedroom. The food Kate insisted on bringing cooled and congealed on the plate, the poached egg horrible as a loose eye, bulbous, viscous.

In truth, Caroline had begun to fade in her memory, like a picture hung for long years on a bright wall. She found and reread her last letter, which she'd never answered. Tortuous, now, what could have been easily done at the time. Happier memories were no less wrenching: Caroline's pink cheeks, her fluttering eyelashes. Dancing around the attic with a stuffed stocking wrapped in a silk handkerchief she made to swish and preach as Madame Praval. Perched amongst the dowdy visitors at Bury Hall on a yellow silk cushion, a canary amongst sparrows. Her breath on Ellen's neck, the pressure of her knee, as they curled together for warmth in the narrow dormitory bed.

She dragged herself up, got dressed, went out. The mountains appeared sharp and clear, yet elusive. The colours she'd use to paint their slopes: two parts lead white, one part blue, one part crimson.

She could describe from memory each step to the summit of Hungry Hill. At first through the bog, a springing carpet of scutch grass, intermingled rock and *Grimmia daviesii*. Sporadic patches of *Campanula rotundifolia, Rhodiola rosea. Thymus serpyllum* var. *hirsutum*, crawling with bees. Levelling out for a spell: a chance to take in the view, the lake a window of cloud and patched blue. Rustling swathes of *Eriophorum angustifolium*. Then the final ascent, heart thumping as she grasped a handful of loose clay, panting, reaching again for a surer hold.

The summit. Two counties spread out, stepped mountains as though echoes, the sea, dazzling, and she, the only human under the sky.

In his crib in the nursery at Bury Hall: Caroline's boy lay blissful, unknowing; wriggling and grabbing for his toes, perfect as shells.

TWENTY-SEVEN

She woke to a commotion – shouting, banging on the front door – and scrambled from bed, heart in her throat. Kate met her on the landing, her face puffed with sleep, mouth agape. They grasped each other, too frightened to move. The banging intensified, like the rattle of hell.

'Damnation, open the door!'

'It's Emanuel,' she choked, relief overtaking shock. They tripped downstairs, nightdresses flapping around their bare ankles. Ellen pulled back the bolts and unlocked the door.

He stood on the step, heavily cloaked, his face a pale slit between his high collar and the brim of his hat. Hard, blowing rain roared into the hall; a storm had whipped up as the household slept. In the driveway, his horse stamped and steamed in the wet.

'For God's sake, what kept you?' he said, pushing past, forcing Ellen against the wall. Kate shrank backwards.

'It's the middle of the night. We weren't expecting you,' Ellen said. 'Why did you not send a message ahead?'

'A gentleman shouldn't have to announce entry to his own house. Get a man out of bed to look after the horse.'

Water poured in rivulets onto the floor. 'Help me,' he said. She put up her arms to take the soaking cape. Her sleeves soaked with rainwater and clung to her gooseflesh skin. 'Not you, Ellen! Go put some clothes on, cover yourself.' She hesitated, unwilling to leave Kate alone. 'Go! Then come back down. I'll want a bath, and food.'

'The fire's been out for hours.'

'Have it lit.'

She and Kate looked at each other. 'Fetch a lad from the stable, Kate, when you're done helping the master.'

Upstairs, Jack called her name. She opened his door. He'd lit a candle; his face glowed, hollowed out by shadow. He sat up and clutched the sheet.

'What's happening?'

'Emanuel is home.'

His mouth fell open. 'What in the blazes is he doing here?'

'I don't know,' she said. 'But his mood is dark. He'll have the entire household in an uproar. Stay in bed. With luck Mother will sleep through. I have to dress.'

Once in her room, her teeth started chattering. A violent burst of coughing bent her double, phlegm propelling from her mouth across the floor. It took a moment for the spasm to pass. Her fingers fumbled with the hooks of her dress, she threw a shawl on top. On the stairs she met Emanuel, his face illuminated by her candle. With the cape and hat removed, she could see the grey stubble on his chin, his bloodshot eyes.

'Send hot water up when it's ready. Bread, meat. And whiskey.' His boots thumped up each step.

'Mother is asleep,' she warned, as he went past. 'And your room is probably damp. It hasn't been aired in a long time.'

'Then I'll need fresh sheets, a fire.' He disappeared into the gloom. Though she could hear him bumping against the walls, he had found his way in the dark.

In the kitchen, Kate kneeled on the flagstones, feeding sticks into the fire. Her hair gleamed with wet.

'Did you get Patrick up?'

'I did. So groggy he insisted he was dreaming, that I was a ghost.'

'I'm sorry for this, Kate.'

'It's not your fault, Miss.' A flame sparked in the dark. The kindling took and blazed to life.

'He wants food, and whiskey.'

'More, you mean,' she said, grimly. 'Lucky not to have been thrown and broken his neck, left in a ditch. In this mood, he'll need to be humoured. Let's get him watered and fed and to his bed. Then we can get back to our own.' She added larger sticks to the fire, expertly stirred the blaze. 'He'll sleep it off until lunchtime.'

'He's insisting on a bath.'

'He must wait. I can't make water boil any faster than nature arranged it.' She sat back on her heels. 'Why has he come now, Miss?'

'I don't know.'

She could not discuss Emanuel's business with Kate. Bound by rules of loyalty, barriers that could not be crossed.

He had come to claim what he thought was his. From those who stood in his way.

TWENTY-EIGHT

'This is where you keep your plants?'

He watched her from the open doorway. She laid the *Byssus barbate* specimen down on the prepared sheet. He had shaved and dressed in clean clothes. He smelled of soap, some kind of pomade. Apart from his red eyelids, he seemed refreshed. They had let him sleep; it was now past noon.

'Yes.'

He pursed his lips and nodded, looking around the room. 'How ordered it all is. You spend much time in here?'

'As much as I can.'

He cleared phlegm from his throat, brushed his hair back from his forehead. 'What a night.'

'You chose a bad one to travel.'

'That's my business.'

'You could have come to harm.'

'Well.' He shrugged. 'I'll admit I didn't anticipate the weather.'

'Did you come from Cork?'

'Initially. I called in on some friends on the way.'

'Near here?'

He folded his arms. 'Near enough. No one you would know.'

'What has brought you back? You didn't write ...'

'I didn't know I needed a reason to come home, sister.' His tone was mild, though his eyes flashed. 'I've come to see how our mother is, for one thing. I was alarmed to hear how she'd deteriorated.'

'Yet you didn't reply to my letters.'

'Oh, I never have time to read those.'

'You don't consider them important?'

He smiled to himself. 'Have you any idea how much correspondence I receive?'

'I can't imagine. Though I would have thought news from home should merit your attention.'

He shrugged. 'Don't worry, I'm well informed. As I need to be.' After a pause, 'Anyway, how is she?'

'I think the worst has passed.'

'I'll see her after I've eaten. She's waited this long.'

She would not lose her temper. 'Have you everything you need?'

'Until the rest of my things arrive, I must make do.'

'Will you stay long?'

'Questions, sister. Too many questions. I haven't decided yet. Nothing compels me to return to England in a hurry.'

She gave up, got to her feet. 'I'll see about preparing you a meal.'

'And have my linens washed. They're filthy from the mud of the road.'

She stopped at the door. 'Try and be civil to Kate. Your coming here brings a lot of work.'

'She'll find me civil enough. As much as a servant deserves.'

Later she peered into his room. Clothes lay strewn across the floor, his stockings hung from the bedpost. A smell of damp persisted even with the window open. She must speak to Kate about lighting a fire.

A week later a coach arrived, laden with trunks. Whatever Emanuel's intentions in Ballylickey, it appeared he had left London for good.

❦

Lettie retreated to her basket and curled in on herself, trembling.

'I don't know what's wrong with the dog. 'Tisn't the month for thunder.' A mound of potato and flour dough flattened and spread under the weight of Kate's rolling pin. 'Not warm enough. The sky doesn't have that iron smell.'

Ellen bent over the basket. 'D'you hear that?' She tickled Lettie's ears. 'There's no need to be frightened.' Lettie sniffed at Ellen's fingers and whined. 'Maybe she smells the cow from when I brought in the milk. She was lucky not to be killed from that kick yesterday.' She clicked her tongue. 'Come on, good girl. We're going for a walk.' Lettie wriggled further into the basket, showed the whites of her eyes. 'All right.' Ellen stood up. 'Stay there, silly creature. You'll demand to go later, and I won't have time.'

'Where's the master?' Kate said. She meant Emanuel, not Jack.

'Somewhere in the fields,' Ellen said. After the first day, Emanuel rose early, almost as early as she did herself, roaming the lands before returning to devour a large breakfast by himself in the dining room. He required clean table linen at every meal, changed his shirt twice a day. Luckily the weather had been fine and the washing dried quickly.

'Should I clean his room?'

'Leave it.' He had ordered that no one enter and cursed the lack of a key for the door. Ellen wasted an afternoon searching in dressers, boxes, trunks and drawers, but a key could not be found. Lost, more than likely by Emanuel himself, years before.

A sharp twinge spiked her hip, the first of the day. She took a walking stick from the corner, took satisfaction from its heft in her palm. Walking loosened the joint, and she needed the stick for navigating crumbling stone walls, the treacherous mud pools that oozed and sucked in the shade, never fully drying out.

'Be careful, Miss,' Kate said, from habit.

'I'm only going as far as Kealkil.'

A bright day. Billowing clouds from the west made even the mountains seem insignificant. She raised her walking stick to the sky, as if they were hers to command.

She walked right into them. Her cursed short-sightedness. Two dark figures, incongruous in the familiar landscape. She kept walking, trying to disguise her limp. Too late she realised it was Emanuel, standing in the middle of the top field with a stranger. The man held a large sheet of paper, while Emanuel strode about, gesturing back towards the house. He joined his companion and stabbed at the paper with his forefinger.

Should she stop and greet them, allow Emanuel to introduce his friend? The unrolled sheet – a map? plans? – had come loose in the stranger's hands and snapped in the breeze, threatening to fly away. Ellen raised her stick in acknowledgement, walked forward to meet the two men. Emanuel stiffened, glared at her like a dog warning a lesser animal away from its meat.

He made himself clear. She turned away.

On the walk back, she noticed a dark object attached to a fence post. It flapped faintly in the breeze. She went closer, curious, then stepped back in horror. Someone – one of the farmhands, or a trespasser? – had nailed a live crow to the wood. Its eye glittered, its beak opened, closed. The flapping came not from the wind, but from the bird's death throes. She stood, helpless, for a moment, before turning away. Bitter bile flooded her mouth. She had not the courage to end its suffering.

Cruel warnings, petty cruelties: the beauty in the ditches, the natural order of the fields and hedgerows, overruled by the mean whims of man.

Emanuel and Jack ate dinner together in the dining room. Ellen hovered between the kitchen and the hall, alert to the low rumblings of conversation beyond the closed door.

'Stand aside, Miss,' Kate hissed, a tray balanced on her hip. 'I need to bring this in.' She opened the door. Emanuel's voice rang out, 'You see my position …'

A minute later, Kate came out.

'Come with me to the kitchen, Miss. Don't stand about the cold hallway.'

'What are they talking about?'

'They stopped when I came in.'

When she went again to take the dirty plates, she reported back that Emanuel had gone to his room. 'Master Jack is still in there.'

Ellen immediately left the kitchen, leaving Kate scraping the leftovers into the pigs' bucket.

Jack sat at the table. She closed the dining room door behind her. 'Has he said anything?' she asked.

He drained his wine glass, wiped his mouth. 'He intends settling here,' he said. 'To breed horses, maybe. He speaks of establishing ownership of the house. And the lands.'

'You know Arthur disagrees with such a plan,' Ellen said.

Jack frowned. 'How do you know that?'

'Did he not talk to you? Emanuel wrote to him.'

'Arthur said naught to me. Anyway, what can he do to stop it? Set the law on him? Emanuel is a Hutchins. And the oldest. At least he's taking an interest in Ballylickey.'

'Can you imagine us all, living together day to day? At the same table, every evening?'

'He'll keep to himself, as he always does. Sam found a way around him and so will we.'

She started pacing the room, twisting her hands. 'I met him in the top field today with a gentleman. No one from around here. I think they were discussing the house.'

'What could that mean?'

'I don't know.' She stopped still in the middle of the room. 'Perhaps he intends to raise money on the lands, somehow. He wants Ballylickey, but not to live in.'

'He couldn't do that.'

'He could try.' She held her hand to her forehead. 'Why is he acting so strangely? He's sat just once with Mother. He avoids me, leaves the room if I walk in. Guards his own den as though he had someone, or something, hidden in there.'

'There's nothing. Just a jumble of papers.'

'How do you know?'

'I went in there, while you were all out this morning.'

Ellen stared at him. 'Jack!'

'Kate helped me upstairs,' he said. 'And stood guard.'

'So you've wondered, too,' she said.

'I admit he seems agitated. He hinted at the old Wolfe Tone trouble … I hope he isn't still mixed up in that business.'

'Surely not.'

'I've heard rumours, over the years.' He drummed his fingers on the tabletop. 'I don't think things went very well for him in London. But we've had a talk. He has ideas for the land. I promised to speak to Arthur on his behalf.'

'Don't get caught in the middle, Jack.'

He sat upright, lifted the wineglass and set it down again with a bang. His mouth quivered. 'I've been master of Ballylickey for years. I have some say in what happens here.'

'I hope so,' Ellen said. 'For all our sakes.' Her face grew hot. Icy fingers played along her spine.

'My God, Ellen. What ails you? Sit. Take some wine.' He filled the glass. To please him, she sipped at it. 'Drink it all,' he said.

Slow warmth spread through her stomach. Her heart ticked and slowed.

'That's better,' Jack said. Relief in his voice, tinged with fear. 'Your skin went green for a moment. Don't make yourself ill. The matter need not concern you, at any rate.'

An echo of Arthur's words, two years before. Authoritative, assured. Yet the twitch at the corner of his mouth showed his true feeling, as it had Arthur's. Subtle. Only those who knew them, loved them, would see it.

'What was that?'

She got up, went to the window. The sky cracked across the bay with jagged light, followed by a release of distant, low crashing, like boulders rolling down a flight of stairs.

'Thunder,' she said. 'And lightning – over the sea for now, but coming closer.'

Lettie had smelled it, when she and Kate couldn't.

Days of unnatural quiet followed, as if the house itself held its breath, broken by a slammed door, a shout for Kate, Patrick the stable boy, Ellen, to fetch something, find something, prepare something. After these outbursts a queasy resettling, like the surface of a pool after a rock had been thrown in. Emanuel spent his time slumped in a chair in the parlour, brooding over letters, or stalking the boundaries of the land, interrogating Jack afterwards on fences, ditches, walls, tenants. Often Ellen had the answers to his questions; he listened without looking her in the eye, staring into the distance. For entire blessed days he rode off, without stating his business or destination – Jack said Cork, though he didn't ride in that direction. Exploring the countryside, she presumed, though where did he stay? Someone put him up, fed him, watered his horse.

One Saturday morning, he took off straight after breakfast. Ellen had slept badly and went to her room; she would try and rest before beginning her letters. Half an hour later, she heard the clatter of hooves on gravel. He must have forgotten something, or changed his plans. She dragged herself off the bed, to the window.

Tom Taylor: dismounting his horse, handing the reins to Patrick, who'd come running.

She stepped quickly back from the window. From behind the drapes she looked again. Yes, it was definitely he – she recognised his stance, his voice. She took off her house apron, smoothed her hair before the mirror and pinched her cheeks. Nothing she could do about the dark circles under her eyes.

She occupied herself for a moment, pointlessly tidying the dresser, scooping away the fallen petals from some roses she'd arranged the day before in a water glass. Wild flowers faded so quickly they were scarcely worth the trouble. She left the room. Coming downstairs, she had the advantage of considering him from above, the patch of dark hair roughened at the back from where he'd pulled off his hat. He ran his hand over the crown.

The step creaked under her weight; he glanced up.

'Here she is.'

Stepping down into the hall, she held out her hand. 'Tom. What a surprise. We're very glad to see you.'

'How long has it been?' he asked, shaking her hand.

'A year, at least.'

'Yes, I believe it is. The honeysuckle was profuse in the hedgerows then, also. But the flies today!' Releasing her hand, he brushed at the sleeves and lapels of his brown coat.

'It's the humidity. We keep the doors and windows closed.'

She led him into the parlour. Sitting down, he pulled a handkerchief from his vest pocket and mopped at his forehead.

'I hoped for fresher conditions by the sea.'

'Not until the wind changes. We've had summer storms, already.' She eased into the chair opposite him, taking care not to wince. 'May I ask after Mrs Taylor? I hope she's well?'

'She stayed in Dublin. She finds the journey tiring, particularly at the moment.' Ellen looked blank. He added, 'She's expecting a child in the autumn.'

'Oh! I see.' After a pause so brief he'd surely not notice, she smiled. 'Understandably, she would not want to travel. How wonderful for you both.'

'How are you all here? First, I should enquire after your mother.'

'She's confined to upstairs. You know she's quite deaf now.'

'That is a burden on you.'

'She is so amiable, and pathetic, that I can't mind.' She pulled at the wristband of her glove. 'Emanuel has returned, which makes her glad.'

'Dr Stokes met him in Dublin. Is Emanuel here now?'

'You've missed him. He left early, to go I don't know where. He might not come back until late, if at all.'

'I had hoped to have a word with him.' He crossed one leg over the other. 'No matter. And how are you, Ellen? What of collecting? Any new or intriguing specimens?'

She shook her head. 'It's been many months since I've ventured beyond a brief walk around our lands. I keep a look out from habit, but it seems my powers of observation have faded, for I find nothing. Mr Turner is surely disappointed in me.'

'Never.'

'All I have to write about these days are my daily troubles and concerns. Which are of little interest to anyone.' A tickling irritation in her throat choked her last words; she grabbed her handkerchief from her sleeve. These days the cough gave no warning, but erupted fully formed into a hacking spasm. As she bent over, pain skewered her hip.

Tom went to the table and poured a glass of water from the jug, setting it beside her. She wiped her eyes, drew a deep, shuddering breath. He laid a hand on her shoulder.

'Are you all right?'

She sipped at the water, put the glass down. 'It's passed.'

'You seemed to be in pain.'

She ignored this. 'Do all physicians do that?'

'What?'

'Look in that way. Dr Stokes did the same. Like I was under his microscope.' She paused. 'It's disconcerting.'

He held up his hands. 'I apologise. I'm unaware that I do it.'

'I suppose I do the same ... search too deeply.' She sat back. Only a trembling in her legs now, the blood easing once more through her veins, her heart steady. 'May I ask ... what you see?'

He made a steeple of his fingers, rested his chin. For a time, he said nothing.

'Tell me, please.'

He dropped his hands, leaned forward. 'Something ails you, terribly. I see it in your eyes, your brow. And more than that ... you've lost weight, and your skin has an alarming pallor. How long have you had that cough?'

She bit her lower lip. 'It's been worse the last six months.'

'Headaches? Fever?'

'At times.'

He frowned. 'What has your physician prescribed?'

'More bleedings, compresses ... they have little effect except to make me even weaker. I had been feeling quite well until ...'

'Until?'

She stood up. The old clock, on the mantel since her father's time, clanged the hour. *The minutes flee, beat by beat.* 'Shall we walk around the garden, to the shore?'

'With pleasure, if you feel up to it.'

A yellow sky, heavy with moisture, clung overhead. Against its pale glow, the grass, the leaves of the trees, throbbed a startling green.

'Take my arm,' Tom said. Halfway down the lawn they instinctively turned back to face the front of the house.

'I have Ballylickey in mind as a template for the house I'm planning to build in Kerry. Though on a bigger scale.'

'I've grown to love it here. The lands, the landscape.'

'Did you not always?'

'I wouldn't allow myself feel comfortable here that first year. I tiptoed around the rooms, as an unwilling guest might in a stranger's house. There have been times it felt like a prison. But now I feel quite tender towards it. I want to protect it as it has protected me.' His arm was solid, warm; she suspected he'd offered it only because he saw how weak she was. 'Talk to me of Dublin, my friends there. Have they forgotten I exist?'

'Not at all. You're mentioned frequently, in the warmest terms. Do you hear from Mackay at all?'

The punctilious Scotsman. His melodious voice, his weakness for whiskey in the morning. 'Not recently. Since my collecting has been curtailed, I fear we have little to discuss. He kept up his letters for many years, from kindness I think.'

'Nonsense. He always admired you greatly.'

And she admired him. Yet something had prevented her from sharing deeper confidences, their letters staying polite, warm but never intimate. As if reading her mind, Tom said, 'And Dawson Turner?'

Her voice brightened. 'His wife had another child. A boy, this time.' More soberly, 'You know two of his sons died, previously?'

'I did not.'

'He never said how the first poor child died. But the second expired as an infant, from cholera. I wonder, could the arrival of a new child erase the loss of another?'

'I doubt it.'

'How anxious he and Mrs Turner must be in the midst of their joy, with all that has gone before. I could not endure such extremes of emotion. My heart is not made for it.'

'We none of us feel equipped for the loss, or pain, we're dealt,' he said. 'We live through it and survive, nonetheless.'

'Yet how are we altered afterwards?'

'That depends on our personalities. Some withstand blows of fortune that would fell another, become even stronger.' They had reached the beach; their feet crunched on the shingle. 'Do you still go out on the boat?' he asked.

'Not this summer. The motion of the waves makes me ill.'

'It gave you so much pleasure.'

'Yes. I fancied myself a veritable queen of the sea at one time, like Grace O'Malley.' She smiled. '*Fuci* for my treasure.'

'You all but conquered the world. One small corner of it, at least.'

'Though there are surely still undiscovered specimens even here, after all my toil.'

'Does that irk you?'

She laughed. 'You know me well.'

Something burst from the water with a loud splash; they turned, but it had disappeared again, leaving only ripples.

'A porpoise?' Tom asked.

'Or a seal. How blessed one feels when they appear, as though they bestowed great favour.'

'Tell me of the rest of your family. Emanuel – I hear he intends on living here at Ballylickey?'

'It appears so,' she said. She withdrew her hand from his arm. 'Where did you hear of it?'

They stopped walking. Tom stared out to sea, as if hoping the creature might reappear. The breeze tugged at his hair. He squinted against it; harder now to read his expression. 'He mentioned something to Dr Stokes, when they spoke in Dublin,' he said.

'Ah.' So Emanuel had confided in someone, outside the family.

'What does it mean for you?' he asked.

An oystercatcher had left its arrow head tracks indented in the firm sand close to the water's edge.

'Ellen?'

'I must wait for my brother's word, I suppose,' she said. 'What else?'

'So he hasn't yet spoken?'

'Not to me.'

He nodded. 'Will you … ask for my help, if you need it?' he said.

'I don't think it will be necessary,' she said.

'Still. Will you?'

'I wouldn't trespass on your kindness.'

'My dear woman.' He shook his head. 'So guarded, so proper. We are old friends, as well as being cousins, are we not?

So tell me, would you trust me to help you, if help were needed? As I fear it might?'

His vehemence unnerved her. 'You seem to know something I don't. You ask me to trust you. Very well – justify my trust by speaking frankly.'

He considered, rubbed the back of his neck. His skin had a yellow tinge, almost sickly in the grey light. 'Your brother wants control of Ballylickey. And would prefer it if you were gone from it. For good.'

For a moment she almost laughed. 'I don't believe it.'

'It's true, I assure you. He told Dr Stokes.'

'Why should he? We haven't opposed him in any way.'

'Who knows? He assumes you will take Arthur's part, and stand in the way of his acquiring Ballylickey, you prove too great a burden on his care … It makes his position easier if you leave. Stokes protested, of course, expressed his abhorrence at any such notion. But Emanuel's a volatile character. Not even his old friend, who he esteems so greatly, could dissuade him.' He held up his hands, anticipating her protest. 'He is your brother, no doubt you feel a great amount of affection towards him. But he doesn't have your interests at heart, I assure you. His intentions, if carried out, would be detrimental to your happiness, if not also your health. Stokes says he's in the grip of some kind of mania. Incoherent with a long list of the grievances, real or imagined, committed against him by Arthur. Stokes wanted to come down himself instantly, or write to you, but I persuaded him to let me go.'

'There must be some misunderstanding,' she said. 'If not on your part, then on Emanuel's.'

'It's unconscionable. He's lost his reason.'

She pulled off her gloves, bit at her thumb, the better to think. 'What will you do?' Tom asked.

'What can I do? Hope you're wrong. Wait, and pray he changes his mind.'

'If not?'

She had no answer, began to walk away from him towards the house. 'Thank you for your honesty. It's grown chilly, I want to go back.'

He followed her, shouting into the breeze. 'I hope this doesn't come to pass, that his words were idle bragging.'

She couldn't hear any more, now.

No moon: the road would soon be dark as pitch. At seven o'clock Tom declared he had to go, though he had hoped to speak with Emanuel.

Ever since their conversation on the beach, she'd willed him to leave. The effort to smile, nod, speak lightly, left her near screaming. 'I thank you for your concern, Tom,' she said. 'Truly, I do.'

'I'll call tomorrow.'

'No.' She pressed his hand. 'I'll write to you at Inchilough.'

He beat his riding crop against his leg, considering. He nodded. 'If you think that's best. I'm a half hour ride away if needed, until I return to Dublin.'

'That is a comfort.'

'You know …' His words trailed away. How unlike him to leave a sentence hanging!

'Yes?'

'It's my regret that I did not – could not – offer you more, when I had the chance.'

Her throat constricted, blocking words. She smiled, flapped her hand stupidly at him, half dismissal, half farewell. After he'd disappeared down the drive, she walked back into the house. In the hallway she leaned for a moment against the wall. Her jaw had spasmed, and she'd bitten her tongue. Her body, as ever, connived to distract her.

TWENTY-NINE

All night, she listened for the creak of the stairs. Morning came and he hadn't returned. She got up, stiff and aching, went downstairs to Jack's room, knocked softly. As she went in, his eyes opened. He hauled himself up, leaned against his pillows. She perched on the edge of the bed.

'Did you sleep?' she asked.

'Not much. The gulls, squawking on the hour.'

'I heard them.' Single frantic calls that clamoured out of the dark, rose to a din, subsided, rose again. 'I wonder what ails them,' she said.

'Excited by something. A dead animal, a rabbit, or a hare maybe. Close to the house, with its guts open.'

'Lettie would have sniffed that out.'

'She'd more likely avoid it, if she caught wind of it. She's no fool, that dog.'

'No.' The springs creaked as she stood. 'I'll bring your breakfast.'

'I'm not hungry.'

'Coffee, at least.'

In the kitchen, Kate stirred the oats while Joanna set out cups and saucers. 'You're up, then, Miss,' Kate said. Her habitual greeting: comfort in the obvious.

'I couldn't sleep. And Jack is awake.'

'This early! He must have stomach ache, or wind.'

'He only wants coffee.'

'Very well.' The long spoon stopped its circling of the pot. 'The master hasn't returned.'

'No, not yet.'

'When do you expect him?'

'When we see him.'

Kate stirred on. Outside, the sound of boots on stone as the yard stirred to life. Chickens squawked and squabbled under the window. Somewhere, Lettie barked.

She took the tray into Jack's room, kicked the door shut behind her. The coffee cup chattered against his teeth. 'No sign of Emanuel?'

'Kate says not. I listened for him in the night.'

'So you haven't slept either. What a pair we make. And then there's Mother, a deaf invalid. Unable to even feed herself.' He leaned his head back, gazed at the ceiling. 'Does it seem like it's all coming to an end?'

'What?'

'Us. Here. Ballylickey.'

'Don't say that.' He spoke as if he already knew. Should she tell him what Tom had said? She decided, no. It might make it real.

'Forgive my mood this morning, sister. This oppressive weather affects my spirits.' He finished his coffee, rattled the cup down. 'What needs to be done today?'

'There are bills to be settled, if you feel like looking through them. You have tenants calling this morning, though I can send them away if you're unwell.'

'No. I'll see them. Refill my cup and I'll make an effort to rouse myself. What will you do?'

'I'll get Mother up. Then a walk, later on. I may do some writing.'

'No drawing?' She shrugged.

He smiled, a hard line that stretched across his teeth. 'See? It affects you too.'

She bent her head. 'So it seems.'

Swiping through the nettles, blood pumping in her ears, she worked herself into a state of forgetting. Defying the pain, the fatigue, the low, miserable sky. The ground grew rockier, rougher, the boreen turned into a track. She stood aside to avoid being trampled by a loose herd of scraggy sheep; they bleated as they scrambled over the stony ground, some with the throatiness of old men, some with the high harangue of young matrons. The shepherd, a scrawny youth with a shock of dark hair, nodded at Ellen as he passed, then yelped at the flock, waving his arms to scare them forward.

The walk grew steeper. She paused, breathing hard, leaned on her stick. The track stretched ahead; she'd easily climbed it only months before. Now it seemed to recede with every step to an impossible distance. If she were to collapse here, only God knew when she'd be found. Reluctantly, after only a couple of hours, she turned back. Her eyes roamed the ditches alongside; in a boggy patch of field she spotted the starburst flowers of *Sium verticillatum*: a rare find. How many times had she walked the road without seeing it? She considered climbing through the ditch; she might never happen upon it anywhere else again. After a moment she went on. It could wait, for another day.

At Coomhola Bridge, the beat of approaching hooves drove her in to the edge of the road. The horse galloped past, the rider cloaked, his head low. Emanuel: she recognised the particular set of his shoulders, hunched inward like a brooding bird. No acknowledgement: either he hadn't seen her or had chosen to ignore her.

Dark spots splattered the ground, and she walked faster.

Close to the house, the sky opened. Rain danced off the surface of the road, throwing up mist. She pulled up her hood and ran, blinded.

Kate came to meet her, holding a blanket over her head.

'The master's back.'

'I know,' Ellen gasped. 'He passed me.'

'Get in, quickly, you're soaked.'

The kitchen was dark as twilight. Ellen pulled off the sodden cloak, shook out her skirts.

'Stand by the fire,' Kate said.

'Where's the master?'

'In his room. I brought him food.'

'How is my mother?'

'Awake, I think.'

Once she'd stopped dripping she went upstairs. Leonora sat beside the window, cocooned in layers, padded out to fill the chair. Ellen had persuaded her to at least move to the window, facing the sea, for a few hours of the day.

Her skin smelled sweet and dusty, like summer grass. She opened her eyes and pawed at Ellen's dress. 'You're wet.'

'Not very.'

'Emanuel came.'

'Yes?'

Her eyes dimmed. She existed in a twilight world of shadow and slumber.

'Mother? What of Emanuel?'

'He said.' Breath. 'Sorry.'

Ellen bent closer. 'He's sorry? For what?'

'Sorry.'

'I know. But what for?'

Leonora's lips moved against Ellen's ear. 'Sorry …'

Ellen stood and looked out at the rain sheeting against the window. As if her mother could hear, she said, 'It's all right. It's all right.'

Emanuel stayed in his room. Kate brought him a tray at dinnertime, took it away an hour later. Ellen waited at the foot of the stairs.

'He says not to disturb him for the rest of the night,' Kate said.

'Show me his plate.' Clean, except for a curling of meat rind and gravy scrapings. Naught wrong with his appetite, then.

Jack had not wanted dinner; when she knocked on his door there was no answer. She called softly as she went in. Only the crest of his hair showed above the bedclothes. The faint rise and fall of his body showed he still breathed.

She ate alone in the parlour, with the ticking clock for company. The evening hadn't brightened after all; at eight she lit the candles and sat down to read Serassi's *Life of Tasso*. She settled for whole minutes, during which she managed a sentence, a paragraph, two paragraphs, before fear began to creep again. The dark subject of her book – the miserable end of the great poet – did little to cheer her.

Would she wait on her fate? Try and reason with him? They could live together, surely, all of them. They'd adapt to Emanuel's presence, find a way to work around each other, maintain a steady if fragile peace. She looked again at the clock, decided. In another fifteen minutes, she'd go.

Ting! The silvery, heartless chime told nine o'clock. She put down the book, took up a candle.

Down the passage, thin light leaked from under the kitchen door – Kate, finishing her last chores. Clashing crockery, muffled voices. She stepped into the dark of the stairway, began to ascend. Decades of Hutchins ancestors regarded her, illuminated by the candle glow.

Whose side were they on? The son and heir? The master of Ballylickey, if and when he chose? Or those without power: the women, the cripple?

The faces were inscrutable, darkened by the dirt of centuries. Could they ever have been truly alive, burned with anger, cherished an animal or a child? Twisted in their beds from passion, or sorrow?

On the landing she paused at Emanuel's door. Light showed through the keyhole. She listened for movement. No sound. Then the floorboards creaked, a weight shifted across the room. He cleared his throat, coughed, sighed. She held her breath.

Swallowing, she lifted her fist, and knocked.

THIRTY

'Who is it?'

'Ellen.'

Scraping on the bare boards. 'What do you want?'

'May I speak with you?'

'It's late, sister.'

'Just for a moment.'

Muttering. Something slammed shut: a drawer. 'Well, if you must.'

She pushed open the door.

He sat at his writing table in his shirtsleeves, elbows bent amidst a mess of papers. More papers pooled around his feet. The candle flickered, casting his face in unsteady light.

'Come in, for God's sake. There's a draught.'

She closed the door and looked for somewhere to sit. Books and cast-off clothes took up the bed. Emanuel leaned back, balancing the chair on its back legs.

'Well? What do you want at this hour?'

'Thomas Taylor visited yesterday.'

He rubbed at his eyes. The rims were already red raw. 'Oh? A bit out of his way, I would have thought.'

'He was on his way to Kerry.'

'I see. And how is Taylor? He might have waited for me.'

'He did wait, until the light began to fade and he could wait no more.'

Emanuel snorted. 'In a hurry to get back to Inchilough and his new wife, I suppose.'

'He came alone.'

'A sacrifice, I'm sure.' He picked up the glass that rested on the table, sniffed the contents. 'Don't just stand there with your mouth open, Ellen, for God's sake. What of Mr Taylor?'

'He had news. Surprising, and alarming.'

His eyes narrowed. 'Well?'

'You intend taking Ballylickey for yourself. And it seems your plans do not include provision for myself or Mother. Or Jack, for all I know.'

His expression remained unchanged. The legs of the chair fell forward with a bang. 'Taylor reported this?' he said. 'From Stokes, I suppose.'

'Yes.'

The hoarse, fussing shrieks of roosting rooks through the open window. 'And what did you think of that?'

'What did I think?'

'Of this information, imparted by Mr Taylor?' He folded his arms and looked at her with casual interest, as if he'd asked her opinion on some abstract matter of politics or farm business.

'Naturally, I found it hard to believe.'

'Indeed?' The corner of his mouth twitched.

'Though he insisted those were your very words, I said there had been some misunderstanding. It makes no sense.'

He shook his head, rubbed at his hands; stage emotion, overly played. Something rattled in the chimney – loosened earth, crumbling masonry. Ellen jerked towards the sound. Emanuel ignored it. He swept a hand across the room, taking in the table of confused papers, the trunks of overflowing documents, the scattered mess on the bed. 'You see how complicated my affairs are? I've been sitting here nights attempting to bring order to it all. Leases, agreements, wills, deeds ... a family as old as ours has a history like a tangled thread. A thread that separates or is pulled over the years into multiple threads, a complex weave. Do you understand?'

She hesitated. 'It must be difficult, I know.'

'Even with my legal training I find it challenging to grasp. But what seems clear enough to me is that I have a right to the title of this house and lands. Despite what others might say.'

She absorbed this for a moment. 'I don't know anything about that, Emanuel. Arthur thinks—'

His face soured. 'I care not a fig what Arthur thinks.'

She sighed, exasperated. 'So you and Arthur disagree. What has it to do with Mother, Jack and I?'

'I have plans for Ballylickey. You would be better off somewhere more … suitable.'

'This is our home.'

He jumped to his feet. The chair crashed to the floor. She stepped back.

'God damn it,' he shouted, 'I'm forty-four. I'm at a time of my life when I need to make plans for the future. I must consider my interests.' He paced back and forth, muttering to himself, before whipping around to face her. 'Listen to me, Ellen.' Attempting reason. 'It's not like you haven't anywhere to go. Arthur will take you in, all of you. He and his saintly wife. Though perhaps you should go elsewhere, escape this place. It's turned you into an old maid.'

'You sent me back here, Emanuel.'

'And now I give you permission to go.'

'You said my duty …'

'Duty!' He spat out the word. 'What's it worth? Nothing, when it stops you getting what's yours by right. I absolve you, sister, of your *duty*. You might as well strive for what you want, the same as the rest of us.' Spit flecked the corners of his mouth.

'So it's true.'

'What?' He wiped his lips, the outburst already subsiding.

'You prefer us to leave.'

Silence. Then, 'You must suit yourselves. But this house will be run from now on as I see fit, for my purposes. If you disagree with that, you're free to go. In fact, I encourage it. Go back to the Stokeses, or your fine friends in England.'

'You put it on us. As if we had a choice.'

'No hysterics, sister. I gave you more credit.'

'What of Mother?'

He picked up the chair, set it aright. 'Her life is all but spent. She'd be better off in town anyway. Bandon, perhaps. Closer to the physicians, and church.'

'Oh, humbug, Emanuel.'

His eyes widened. 'What did you say?'

'You may justify your behaviour any way you wish. The truth is you care not a whit for Mother, or anyone but yourself.' She headed for the door. 'Arthur shall hear of it.'

He came across the room and grabbed her arm. She tried to twist away. 'If Arthur comes here, he'll be sent packing. With force, if needs be.' They now stood nose to nose. She could see the individual wayward hairs in his eyebrows, the circle of amber around his irises.

'What is that?' He flicked at her chest.

She leaned away. 'What?'

'There.'

She groped around her neck, touched the chain. It had slipped into view. He pushed her hand away and pulled out the locket.

'Who gave you this?'

'Jack.'

'By whose permission?'

'He thought I might like to have it.'

His eyes had hardened to pale blue stones. 'He had no right.'

'He meant no harm.'

He let the locket fall back against her bosom. 'Jack is too quick to give away my property.'

'She was my sister, Emanuel.'

'You were a child, you could not have known her.' His expression was a mix of pity and distaste. 'Nor do I know you,

Ellen, not really. Isn't that the truth? There were those I cared about—' He stopped, shrugged. 'All long gone.' He lifted the glass from the table, gulped down the contents. 'I'll not take trifles away from you. I'm not that petty. Now leave me.'

'Will you come downstairs?' She tried to keep her voice calm. 'We can sit together, resolve things.'

'There's nothing left to discuss. I wish you good night.'

She stood there for a moment longer, but he had sat again at the table and bent his head to his papers. She moved to the door. Her limbs felt as though someone else controlled them.

'And get Kate to send up more of this.' Without looking at her, he held up the empty glass.

Next morning his door lay wide open, as though he no longer had anything to hide, his sheets pulled back, the bed empty. He had left already, to go who knew where. She ate a solitary breakfast, brooded over her tea awhile. Then she roused herself and went upstairs to her writing desk, took up her quill. Finishing the note with a last plea – *come talk to him, Arthur, I pray* – she sealed it and called for Dan, asked him to deliver it directly, to interrupt her brother no matter what else he might be engaged in. She watched, twisting her hands, as he saddled the horse with maddening care. How slow he was! Eventually he mounted and, clicking the mare into a trot, raised his hand in farewell.

At that moment, a pack of men marched through the front gates. Well-dressed, blank-faced strangers, carrying sticks; in a smooth, organised movement they spread out, blocking Dan's way. He tried to drive the horse through; in the kerfuffle a tall man with a thick, dark beard grabbed the reins. 'Get down, old man,' he said, 'or we'll knock you down.'

Emanuel appeared behind them, on horseback, his face wrapped against the morning air.

YARMOUTH, ENGLAND
AUGUST 1813

THIRTY-ONE

The itch just above his left ankle started again. He tried to ignore it, went back to reading his letter. For thirty seconds he forgot it entirely, before it crept back. He stretched his arm. Could he reach …? No, useless. He fell back into the chair.

'Maria,' he said.

'Yes, Papa?'

'Come here and scratch my ankle. This itch will drive me to distraction.'

Ever since turning sixteen Maria wore an expression of carefully composed gravity. She got up and came to stand by the stool where his foot rested.

'Here, Pa?' She prodded around the ankle, searching for the spot.

'Good girl.' Relief.

'Have you heard from your friend? It's many weeks since the accident.' Her face was now pinched with studied concern.

Dawson's friend had been thrown clear of the coach, cracking his skull on a rock as he landed.

'He's regained consciousness at last. His wife writes that he'll make a full recovery.' Whether or not he'd walk, or talk, as before was a different matter.

'Will you visit him? I could make him a drawing.'

He pointed to his leg. 'I cannot go anywhere until this is healed.'

'Is it improving?' She looked down at the injured limb, which was patched in blue and purple bruises that had just begun to yellow around the edges.

'Slowly.'

'I am glad.' She glided back to her chair, head erect as though balancing a stack of books, and sat, arranging her dress. Did he detect a note of condescension? He smiled to himself. No longer a child, the innocence of only a few years ago fading. Yet beneath her affected smoothness she surely suffered the pangs of change. He must show her particular attention. Fatherhood, an ever-changing challenge: to remain adaptive. Sensing his attentions, she smiled suddenly at him – the familiar smile that crinkled her nose and the corners of her eyes. The little girl remained, in part, at least for now.

Thank God.

'Stay close,' he mock-commanded. 'I may need your services again, Miss.'

❧

Unnamed horror conspired to torment him. Man or beast? Undefined, malevolent ... He shuddered, tried to shake off its grip. Shadows lurked even here, in the dark corners of his pleasant drawing room. He hauled himself up and sucked in a breath, feeling saliva on his chin.

'What's wrong?' Mary asked, from the chair opposite.

'A bad dream,' he said.

'From the laudanum?'

He still needed the reddish-brown drops for the pain. Perhaps he'd overdone it once or twice, reached for the bottle too hastily. He slipped into sleep as though into a bath of warm honey, his dreams had become extraordinarily vivid and lingered well beyond waking. Pain danced on the horizon, never penetrating too far; even his worries about the bank and how it functioned downstairs without him had softened to a sweet *laissez-faire*.

'Take care not to become too attached to its comforts,' Mary said. 'Though I must say, it has helped keep you quiet the last weeks, to the benefit of the rest of us. I have never known you so … static. I should add a drop to your coffee in the morning.'

'I am far from sedated,' he said. 'I read entirely through the morning's correspondence.'

'Anything of interest?'

'Not particularly. Except—' He grasped the arms of the chair and pulled himself upright, wincing. The unease that had lingered after his dream spiked again.

'What is it?'

'Our Miss Hutchins.'

'What of her?'

'She has left Ballylickey House. Quite suddenly, it appears. She now resides in Bandon, a large town some miles away.'

'Does she say why?'

He folded his hands across his stomach. 'To consult with the more experienced physicians there about her mother's health. She describes it as if they are engaged on a pleasure trip, visiting, touring around.'

'It that a matter for concern?'

'She never mentioned such a plan in previous letters. I formed the impression she was deeply attached to Ballylickey.'

'But if her mother requires treatment …'

'Her mother is aged, at death's door now for some time, if what I've interpreted from her daughter's references is correct. Why move her now, to an inn of all places? There's something more at play here.'

'Such as?'

'I know not.' He frowned. 'But her letter … vibrates, with anxiety. Trouble of the mind, as she describes it. What could cause such unease to a gentlewoman living in seclusion, a placid life without dramatic events to disturb it?'

'You identified it yourself. She fears for her mother's health, anticipates her death. From what you say, they're deeply attached to one another. The only other woman in the family – without her she will be surrounded by male relatives. They can't offer the same quality of confidence, or sympathy. She will be quite lost.'

She paused in her mending of one of the children's little coats, holding it to the light, considering, probing at it again with her needle. Her face and hands were still swollen from her most recent pregnancy, lines of fatigue etched around her eyes. He noticed paint on her wrist, a thumb-sized dab of vermillion. He could never look at her without seeing the young woman he'd first married: slender, soft, her face open as a rose. She'd revealed her fierceness, her sharp intelligence over their first months of marriage. Now her face and her character had settled into mature equilibrium. Powdery dust on her cheeks, the first grey at her temples.

'Besides,' she went on, unaware of his gaze, 'as you have described her to me, she is skittish, as intelligent, creative women in her position often are. No doubt whatever distresses her is not so bleak or insurmountable as she imagines. Write and tell her so.'

No, there had to be more to it. Ellen Hutchins's letters now had urgency, a simple expressiveness that proclaimed what? Desperation? Or, as Mary said, a combination of ill health and mania?

'She's on the edge of something,' he said. 'She complains of physical ailments, headaches and the like. But somehow I have the feeling these are manifestations of something else. Some trouble that oppresses her greatly. I could write to Dr Taylor and ask him if he knows anything. Or Dr Stokes.'

'Oh, no.' Mary looked up from her stitching. 'That would surely be a betrayal of confidence.'

'Stokes is like a father to her.'

'Precisely why she might have confided in you, as someone outside her Irish circle.' Mary snipped a piece of thread with a tiny, shining set of scissors. 'Why not invite her again to come here?'

'Or I could go to Ireland,' he said. He'd threatened to do so for years, though in truth, these had only ever been vague musings on his part. Another scheme, more compelling, had always distracted him, occupied his time and energy. Well, so why not now? Invent the pretext of a botanising expedition, call in on Ballylickey House. 'I'll need a change after this damn accident.'

'You won't be fit until the end of the summer,' she said calmly. She cut a last thread, shook out the coat, folded it back into the hamper. 'Not enough to withstand travelling to Ireland, at any rate. Once you're back on your feet, you'll be busier than ever downstairs. And haven't you already committed to visiting Sir James as soon as you possibly can?'

Blast, he'd forgotten. He'd given his word to the old man.

'That's that, then.' She stood up, came across to kiss his forehead. 'You can't heal every broken bird, Mr Turner. Not before you heal yourself.'

'All the more reason to write to Stokes.'

'Perhaps. But do ask her to come here. The baby would cheer her. As he cheers us all.' The baby: Gurney Turner. Mary had got her way. Christened in the bedroom with the blood of his delivery still on the sheets, named for Dawson's business partner.

Again, his eyes grew heavy. There might yet be another little Dawson, another namesake. All wrongs righted, in due course. Torn cloth stitched, bones mended.

BANDON, CO. CORK
OCTOBER 1813

THIRTY-TWO

At Ballylickey, voices wove into the silence, natural as the wind. Here in the town, they snagged on the ear, strange and jarring. The weather was close and she left the window open all day. Shouted greetings, chatter, a boy's bright whistle. From the guests in the inn hallway – travelling merchants, or Englishmen touring the countryside, as Mr Dillwyn, Mr Leach and Mr Woods did – sounds of scuffling feet, stifled laughter and coughs. Whispering, treading lightly, for the sake of the ailing woman and her withered daughter in Room Four. They stood respectfully aside when they met the physician on the stairs.

For weeks she'd been forcing down whey, into her own mouth and Leonora's, from a spoon, with a little dry biscuit. She'd grown to hate its clear, milky sourness. The body couldn't function on light and water, as plants did. If she were still capable, she would have climbed Hungry Hill and lain under the clean sky: absorbing, absorbing. Remove her clothes, the better to let the light and rain penetrate her skin: limbs, breasts, stomach, all white as paper, though her hands and face were faintly freckled.

Impossible to collect the tiny plants wearing bonnet and gloves.

Her ribs now protruded so that she could count them. They reminded her of the railings at Platanus; the little girls trailed a stick across them to make a kind of clanging music.

These days of bleak existence, sudden, unbidden thoughts of Madame Praval stole into her mind. Peculiar, when she hadn't thought of her in years. Her piercing black eyes, that seemed to see beneath the skin. Her words of warning, that last evening: *You will wear yourself out, you will fade to nothing, Mademoiselle. Quietly, politely. And no one will notice.*

Yes, she had known her, better than her mother. Though in part at least, the crucial part, Madame had been wrong. Ellen had made sure of it. Her name was noted, beyond Ballylickey, beyond the confines of place. With every year, the plants that bore her name would sprout shoots, stretch forth, from the dank soil to the crystalline light.

A blackbird, she recalled, had sung, outside the dormitory window.

Music, all her life, had been rare and golden; she had scarcely heard an instrument since the evenings when Louise played the pianoforte at Harcourt Street. Here in town, the common people played fiddles and whistles – she'd overheard snatches, floating through some cottage window. The men drank whiskey from eggshells, sang and howled, the women stamped their feet upon the earthen floor. She had no place amongst them, and hurried past. Jack had hummed, sometimes, songs he'd learned at school before he slipped on the ice and broke his back. She herself had no singing voice, but then, she'd noticed it was often the tuneless that most craved the song.

Gulls followed the Bandon River to the town and screeched over the dunghills. When well again, she would trace the same river, back to its mouth where it entered the sea. Home.

THIRTY-THREE

At two o'clock, a knock at the door: Arthur, dressed in mourning black, holding his hat, stooping under the low lintel.

'You're ready, I see.'

'There wasn't much to pack.'

Looking around the bare room, he said, 'No.' His eyes turned to Ellen, roved over her face. 'How are you feeling?'

'A little better. I slept last night, when I thought I wouldn't.'

'Hmm.' He seemed unconvinced.

'Perhaps now that Mother—' She paused, swallowed, went on. 'Perhaps now that my duties have eased, my own health will improve.'

'I've spoken to a physician in Cork, recommended to me as exceptionally capable. He suggests a particular treatment, a tonic – the administering of mercury. He claims it will help you put on weight.'

'I see.'

'You sound unsure.'

'Such treatments have not helped in the past.'

'Remain open-minded, Ellen, give the man your trust. I believe he speaks from sound scientific principle.'

'If you think it best, Arthur. What would truly bring comfort to my mind and body cannot be.'

The lines around his mouth tightened. 'Let's not speak of that.'

She sighed. 'No. You're right.'

'Come,' he said. 'Matilda is waiting.' She hooked his proffered arm, though she could manage the stairs well enough now. During the last days of Leonora's decline, some of her own strength had come back, enough to sit out the vigil at the bedside, tend to her mother's few needs herself. Wet her cracked lips with a cloth dipped in water, hold the bedpan underneath the bird-like body, chivy the servant.

Arthur's servant came in and hoisted Ellen's trunk on his shoulder. She'd tucked away Leonora's prayer book, shawl and silver pillbox. She wore her mother's old wedding ring on her own right hand, Katherine's locket over her dress. The bed linen and clothing had been given to the housekeeper with the instruction to use, sell or burn, as she saw fit. Leonora's crockery, furniture (including the heavy mahogany marriage bed, originally brought with her from her childhood home in Limerick on its own cart pulled by two horses), silver, glassware, pictures – all remained at Ballylickey. As did Ellen's drawings, her herbarium, microscope, drying box, paints. No time that day to pack anything but a few essential items. Inconceivable at any rate to think of any of it outside the walls of Ballylickey.

They'd left it all behind, in the hope of returning. If Emanuel was persuaded to leave, by one means or another. If he simply changed his mind, or his behaviour.

This last option seemed least likely.

The intention had been that the inn would serve as temporary asylum until Arthur could secure Ballylickey again. 'If you come to us at Ardnagashel, that will be the end of any hope of getting him out,' Arthur said. 'We must shame him in front of the county.' But Emanuel proved to be shameless. His mother lay dying in the rented room, with the smell of grease and cabbage drifting upwards from the kitchens; Ellen's only comfort had been that she was oblivious to her surroundings at the end. Leonora fancied herself in the parlour at Ballylickey, calling for Jack, muttering for Kate to count the cutlery, count the cutlery … where was the silver spoon? Her hands clawed at the sheet, her head twisted from side to side.

Kate. Watching hollow-eyed from the door, Lettie in her arms, as the carriage pulled away. Raising her hand at the last moment in a stiff little wave. Less a farewell, more a benediction. Or acknowledgement of something. She had since left, Arthur had heard. The shutters of the house stayed bolted for days on end.

Not even this image could rekindle despair, or rage. Emotion had been leached away. Ellen's energy went now towards existing, breathing from one second to the next. The worn floorboards, the hard mattress on the low cot by the window, the cracked chamber pot and flecked, warped mirror – all adequate to her needs.

Nothing much changed within an hour that could be generally perceived. A slant of sun moved across the wall, illuminating motes of dust. Below on the street, two women hailed each other in Irish, gossiped, laughed, and parted. She knew enough of their language to recognise the words: *Slán leat*. Good health to you. A blessing.

Lie down on the cot. Rest. Rise. Walk slowly from one wall to another. Repeat. Nothing stupendous occurred. Her heart beat on; her skin flaked into dust.

In the carriage Matilda took her hand without speaking. Language had ossified into a dead, useless thing.

Dappled sunlight played about their knees. The road swung close to Ballylickey; they passed the turnoff, left it behind. She stared ahead.

One morning the household woke to the sound of pistol shot. Emanuel, still in his nightshirt, stood at his open bedroom window, firing at magpies on the shed roof opposite. As they ran in he turned and fired into the ceiling, then laughed at their shocked faces. She began packing before breakfast.

Jack had sat in his chair as though rooted there, gripping the arms. His knuckles showed white. Emanuel needed him, to talk to the tenants, to maintain the illusion of normality.

'You understand, Ellen? I can't leave here?'

'I know, Jack.'

For his sake, she swallowed the gush of rage and grief.

For his sake, she promised to return when the situation calmed.

For his sake, she smiled and waved from the carriage: fare-well.

On the seat opposite, Arthur coughed and cleared his throat. Ellen's fingers, entwined with Matilda's, grew clammy, yet she held on. 'Matilda.'

'Yes, my dear?' Her voice was eager.

'May I entrust you with something?'

'Of course.'

'If anything should ever happen to me ...'

'Please don't, Ellen.'

'... Hush. I must. If anything should ever happen to me, I want you to make sure my herbarium and drawings go to Mr Turner of Yarmouth.'

Arthur leaned across. 'This is not the time or place to discuss that, Ellen. Besides, it may not be possible to get them back. You must allow for the possibility they are gone for good. When you're well, you can begin collecting again.'

'Don't treat me like a child, Arthur.'

He opened his mouth, closed it again, slumped back in his seat.

Ellen looked into Matilda's eyes. 'Promise me, sister.'

Matilda nodded. 'It will be done. If I have to go in there myself. Though naught will happen to you, Ellen. We shan't allow it.'

The winding coast road to Ardnagashel: Bantry Bay to their left. A short time later they drew up in front of the house. Once the carriage stopped, they were enveloped by silence. Then, somewhere out of view, children's voices: a high, gay stream on the breeze. At least they continued to grow up here, not yet sent away to school. A gap cleared in Ellen's thoughts, like a chink of light through a black wall. What would little Ellen Turner become? Would she skip on the shore, in a white dress? Shells in her pocket, making music as she walked? Search the beach with her father as a light wind lifted her hair …

Margaret appeared at the window. She'd grown tall, more boisterous, no doubt from having a brother as a playmate. Matilda leaned across. 'Where is your bonnet, miss?'

'I ran,' she said, 'and forgot it in my haste.'

'And when you're freckled as an egg, what will we do with you then?' Matilda whispered aside to Ellen, 'We cannot keep her indoors, the child is wild. My looks and Hutchins's ways …'

Arthur helped Ellen down. Stepping onto the gravel, she took off her glove and held out her hand to her niece. The child's hand was warm, oily and plump. 'Can I do anything for you, Aunt?' she said, instantly shy.

'Do you remember I once showed you a moss? Can you recall what it looks like?'

The child nodded doubtfully. 'I think so.'

'Go into the woods. Find a moss. Bring it to me.'

'Why, Aunt?'

'To look at. I begin to forget.'

Margaret smiled, a child's acceptance of the strange habits of adults, and relatives in particular. 'I understand.'

YARMOUTH, ENGLAND
MAY 1816

THIRTY-FOUR

The crates arrived and were carried laboriously up the stairs with much exhorting to take care, not to drop them, nor bang them against the walls, to set them down slowly, not there, here, in the middle of the study floor: gently, gently, please …

Dawson allowed the men loosen the nails from the lids with a crowbar, but bid them to leave them sitting on top, as though still closed.

One of those sweet May mornings, fair and fresh, winter gone, the dull breath of summer not yet upon them with a stink from the Yare and the endless fretting, sweating, yearning to escape the bank into the countryside. The breeze stirred the drapes, tickled at the papers on his desk. He sat down, took out the letter to Hooker and reread the last lines he'd written.

I just received her herbarium; now I must decide what to do with it … such a legacy, the result of years of toil and learning, invaluable to the history of botany as its devoted originator was to her friends and family, deserves the utmost care. That she thought of me in her final days touches me greatly …

He threw down the quill, spilling ink across his carefully composed words; cursing, he hastily blotted the splatters away. Ruined. He tore the page in two.

Outside, sunlight on water. Clusters of shouting men unloaded a frigate from France: wine barrels, rolled down a plank, were hoisted onto a cart; mail sacks, limp, lumpen as though containing dead bodies, tossed with a thud onto the wharf.

Glib, easy words. How they flowed, shoring up the lies one told oneself. Instead of expressing true feeling.

Guilt.

How sincere had his invitations been? He could answer this at least, look without cringing into his heart: they had been heartfelt, open, genuine, utterly supported by Mary. He asked Ellen Hutchins to come. She would not. She could not. And even if she had broken free, escaped across the sea, found refuge in his house, would it have changed how things ended? Possibly, probably not. And yet ... he would have given her that chance.

What else could he have done? He could have gone to Ireland himself, proved his esteem. Coach to ship to coach, days of dreary travelling. Offered her comfort, clasped her hand one time. He could have brought his daughter Maria – now she was grown up, married to Hooker, he wouldn't have the opportunity to travel with her alone again – shown her what it was possible for a woman to achieve, a woman of limited education, imperfect circumstance, poor health. He had done none of this. He had good reason, at the time. Too late now.

Too late.

Her cousin, Dr Taylor, had written – could it really be almost a year ago? – anger palpable in his stark description of her last months. He laid the blame squarely at the feet of the brother, who he described as a madman. The strength of his loyalty, his passion, surprised Dawson.

Dr Stokes wrote also, thanking him for his support of his some-time ward.

I saw death in her eyes, when she was but eighteen years old. She pushed it aside with all her might, seized the life force. For the next twelve years she lived with its cold breath on her neck, until her spirit failed. To wish she might have had more is futile. She defied fate to achieve what few would even attempt.

Her Irish friends were heartbroken. *My wife*, Stokes wrote, *is inconsolable.*

A tap on the door. Mary, in a violet dress. Unusually for the time of day, she carried the baby. Three-year old Gurney hid behind her skirts. Seeing his father he ran into the room and scrambled beneath the desk, a favourite hiding place.

Dawson stood, glad of the distraction, and went to his wife. He pulled at the baby's fist.

'Well, sir. And why are you not in the nursery?'

'He had a bad night. The girl couldn't settle him. I said I'd walk him about for an hour.' She nodded at the three crates on the floor. 'You haven't looked through them yet?'

'No,' he said. 'Later.'

She nodded. 'A full year, to get the herbarium here.'

'It's a miracle it wasn't burnt in the mayhem. You know the brothers fought over that house like feudal lords over a castle.'

'Shameful.' Mary sighed. 'And Miss Hutchins in the middle of it all.'

'Hard to imagine civilised men acting so. A strange country, certainly.' He frowned. 'You know, I believe I have almost exhausted my interest in botany.'

Her eyes widened. 'You surprise me. What would you do instead? Take up billiards, or fishing?'

'Does that seem likely? But there are many other pursuits worthy of my time.'

'Such as?'

'That church we visited last month. The carvings, the stonework. It struck me that no one has made a record of the ancient treasures of Norfolk, that I'm aware of.'

'I anticipate a new enthusiasm, husband. There's no other reason for your cooling towards botany?'

He shrugged. 'I am already well past middle age. And the greater my years, the more wondrous the world seems. As if I will only ever experience a fraction of God's magnificence before the end. I feel a great urgency to ... *change*. To seize another subject, lose myself in a new area of discovery.'

'You have always felt that impetus.' She laid her hand on his arm. 'Such thoughts only serve to confuse, in my opinion. The world will wait. Better to concentrate on the *now*, lest you lose your sense of place altogether.'

The baby made a sudden sound, as if trying to win their attention. His eyes twinkled up at his father. Clear depths of innocence. 'I perceive you are very far from sleep, mischief,' Dawson said.

'At least he's stopped grizzling. Poor creature. I think more teeth are coming.' Dawson offered his finger. His son clamped his gums upon it, spittle running down his chin. Dawson had a sudden image of his own dying father, given a piece of honeycomb to suck when he could no longer take any other sustenance. The old man's eyes had similarly gazed upwards at his son, full of blank light. Beyond awareness, conscious only of the sweetness on his lips.

He placed his free hand on the baby's skull. It emanated a tender heat. 'There,' he crooned. 'Isn't that better, little Dawson? Yes, that's better.'

Epilogue

The bay lay still and clear, the line against the horizon seamlessly blending into blue-silver.

Slicing through the water, the boat glided into the inlet, a refuge between two large, algae-covered slabs, placed there, surely, since time began, for the one purpose of guiding them in on this day, this perfect day of late-summer. As the boat docked, its bottom scraped against the shingle. She stepped out, climbed over the smaller rocks around the base of the island – itself no more than a large rock – found her footing, scrambled higher. Took a moment to look around, a full turn that took in the curve of the bay, open sea.

The father and his son had stayed in the boat. Their voices drifted to her now, speaking in their own language, melodic and burred. She understood not a word but felt no separation, only kinship. That of the sea, and place. They spoke, she knew, of fish, of tides, of weather and seasons. Family, sons and daughters, events that marked the passing of years. Children, and animals. All these she understood at last, and loved, as they did. They had no resentment of her, or her language. She was part of things, of them, as the mountains or land, the wet, green fields and barren hills.

Then she saw it. Less of a tree, more a large shrub, scrubby, thin limbed. Scattered leaves, and miraculously, flowers. He had not mentioned flowers. Glorious, pink-purple trumpets, with an indigo centre, fathomless as a bog pool. It stood apart, self-contained, occupied only with its own existence. How to withstand the years, surrender but not succumb to the inevitability of gales, the great waves, the salt-laden sea-spray.

Roots deep within rock, tenuous and fierce. She took her magnifying glass from her pocket, raised it to her eye, looked into the heart of the flower.

HORTUS KEWENSIS;

OR,

A CATALOGUE

OF

THE PLANTS

CULTIVATED IN

THE ROYAL BOTANIC GARDEN AT KEW,

BY THE LATE WILLIAM AITON.

———

THE SECOND EDITION ENLARGED
BY WILLIAM TOWNSEND AITON,
GARDENER TO HIS MAJESTY.

———

VOL. IV.

LONDON:
PRINTED FOR
LONGMAN, HURST, REES, ORME, AND BROWN,
PATERNOSTER ROW.
1812.

TETRADYNAMIA SILICULOSA.

HUTCHINSIA.

Silicula elliptica integra: valvis navicularibus ; apteris ; loculis dispermis. *Filamenta* edentula.

rotundifolia. 1. H. foliis inferioribus orbiculato-ovatis petiolatis;
superioribus cordato-sagittatis integris.
Iberis rotundifolia. *Willden. sp. pl.* 3. *p.* 434.
Lcpidium rotundifolium. *Alllion. pedem.* 1. *p.* 252. *t.* 55. *f.* 2.
Round-leaved Hutchinsia.
Nat. of Switzerland and Piedmont.
Cult. before 1759, by Mr Ph. Miller. *Mill. dict. ed.*
7. Iberis 7.
Fl. May – July.

alpina. 2. H. foliis pinnatis integerrimis glabris, petalis calyce deciduo
duplo longioribus, siliculis utrinque acutis; stylo brevissimo exserto.
Lepidium alpinum. *Willden. sp. pl,* 3. *p.* 433.
Jacqu. austr. 2. *p.* 23. *t.* 137.
Alpine Hutchinsia.
Nat. of the Alps of Germany, Switzerland, and Italy.
Introd. 1775, by the Doctors Pitcairn and Fothergill.
Fl. April – June.

petraea.   3. II. foliis pinnatis integerrimis, petalis calyce vix
longioribus, siliculus utrinque obtusis ;
stigmate sessili.

Lepidium petraeum. *Willden. sp. pl,* 3. *p.* 434.

Engl. bot. 111.

Rock Huchinsia.

Nat. of England.

Fl. March – May.

Afterword

Ellen Hutchins died on 9 February 1815, just before her thirtieth birthday. She collected over a thousand plants in the area around Bantry Bay, Cork and completed hundreds of botanical drawings. Her specimens and drawings feature in botanical collections in the Royal Botanic Gardens, Kew, the Natural History Museum, London, New York Botanical Garden, Trinity College, Dublin and Museums Sheffield. Her correspondence with Dawson Turner lasted from 1807 to 1814. The genus Hutchinsia (Brassicaceae) was named for her; *Hornungia alpina* is still commonly known as '*Hutchinsia*' in the UK.

Dawson Turner eventually gave up on botany and took up a new enthusiasm, antiquities, travelling to Normandy with John Sell Cotman, the watercolour artist, who taught Mary Turner and several of the Turner children to paint. Dawson Turner died on the 21st of June, 1858. Mary Palgrave Turner died in 1850. Many of her etchings of the Turner family and well-known figures of the day are in the collection of the British National Portrait Gallery. Maria Dawson Turner (1797–1872) married her father's friend, William Jackson Hooker, in 1815. William served as Director of the Royal Botanic Gardens Kew, as did his and Maria's son, Joseph.

Dr Whitley Stokes's son, William Stokes (1804–1878), became one of Ireland's most eminent physicians; the conditions Chyne-Stokes breathing and Stokes-Adams syndrome are named after him.

Emanuel Hutchins stayed in West Cork and became a magistrate. He died in 1839, aged seventy, in Damascus, while buying Arabian horses for Lord Kenmare.

Thomas Taylor continued to concentrate on botany as well as being an esteemed physician and magistrate. He eventually settled at Dunkerron, near Kenmare. He was consultant to the fever hospital in Kenmare during the Great Famine, and died, himself of fever, in February 1848.

Sources and Acknowledgements

This is a work of fiction. Generally, however, I have kept to the main details known or supposed about Ellen's life and her circle. On occasion, dates, the chronology of events, details and names have been altered in the interests of storytelling and clarity. I have invented all dialogue and letters between characters.

For inspiring this novel, grateful thanks and acknowledgement:

Hutchins, Alice (Ellen's great-niece), 'Ellen Hutchins, A Botanist', Ms 47 (Representative Church Body Library, Dublin, 1914).

Lyne, Gerard J., 'Lewis Dillwyn's visit to Waterford, Cork and Tipperary in 1809' in D. Ó Murchadha (ed.), *Journal of the Cork Archaeological and Historical Society,* 9 (1986), 86–104.

Lyne, Gerard J. and M.E. Mitchell, 'A scientific tour through Munster: the travels of Joseph Woods, architect and botanist, in 1809', *North Munster Antiquarian Journal* 27 (1985), 15–61.

Mitchell, Professor Michael (ed.), *Early Observations on the Flora of South-West Ireland: Selected Letters of Ellen Hutchins and Dawson Turner 1807–1814*, Occasional Papers 12 (National Botanic Gardens, Glasnevin, Dublin, 1999).

Ellen's nature notes are taken from William Withering's *A Systematic Arrangement of British Plants.*

Thank you Aoife K. Walsh and New Island Books for making this book possible and Susan McKeever for her perceptive editing.

Thanks to: the Irish Writers Centre, Cabra Public Library, the National Library of Ireland, UCC Boole Library, the Library of Kew Gardens, London, National Museum of Ireland, Collins Barracks and the many institutions that digitize eighteenth- and nineteenth-century books and periodicals and make them available online.

Máire Kennedy and her website mairekennedybooks. wordpress.com for details on Charles and Catherine Praval and other aspects of early nineteenth-century Dublin literary life.

Any errors in this book are my own. However, I would like to thank Clare Heardman of the National Parks and Wildlife Service, who read an early version of the manuscript and suggested appropriate birds and plants for specific environments.

Thank you to Ted L. Hickey, formerly of the School of Biological, Earth and Environmental Sciences, UCC.

Madeline Hutchins, Ellen's great-great grandniece, and Samuel's great-great granddaughter, keeps Ellen's legacy alive in West Cork and beyond, and has spent much time deciphering Ellen's letters, to the benefit of all who are interested in botany and early nineteenth-century life in South-West Ireland. She shared her insights into Ellen's life and work with me and I owe her particular thanks. www.ellenhutchins.com

The small tree/shrub Ellen searches for may be *Malva arborea*, which does grow on some of the islets of Bantry Bay. This is the plant described in the epilogue.